10 FEB 1996   1 6 MAR 2002   338.

- 2 APR 1996   272.

2 1 JUN 1996   - 1 FEB 2003   THS

I 4 SEP 1996   - 3 MAY 2003

1 4 MAR 1997   09/03

1 8 OCT 1997   2 3 OCT 2003

3 0 MAR 1998   26 JAN 2009

1 3 OCT 2000   4/12

Please renew/return this item by the last date shown.

So that your telephone call is charged at local rate, please call the numbers as set out below:

	From Area codes 01923 or 0208:	From the rest of Herts:
Renewals:	01923 471373	01438 737373
Enquiries:	01923 471333	01438 737333
Minicom:	01923 471599	01438 737599

L32b

H40 857 755 9

Images of Industry
COAL

ROYAL
COMMISSION
ON THE HISTORICAL
MONUMENTS
OF ENGLAND

Images of Industry

COAL

Robin Thornes

ROYAL COMMISSION ON THE HISTORICAL MONUMENTS OF ENGLAND

Published by the Royal Commission on the Historical Monuments of England,
National Monuments Record Centre, Kemble Drive, Swindon SN2 2GZ

© RCHME Crown copyright 1994

First published 1994

ISBN 1 873592 23 X

British Library Cataloguing in Publication Data
A CIP catalogue record for this book is available from the British Library

Designed by John Mitchell

Typeset by Nene Phototypesetters Ltd, Northampton
Printed in Great Britain by BAS Printers, Over Wallop

Contents

Illustrations

Commissioners

Chairman's Foreword

The future of the coal industry and its associated sites and settlements has been the subject of intense debate in recent years. Aspects of the story of the industry, which is closely intertwined with that of the industrialisation of the country as a whole, are related here in photographs of the structures, which owe their existence to coalmining, and the people and settlements associated with it. The text and photographs, all of which were taken by RCHME photographers during the last five years, present both the familiar and the unexpected. They do not attempt to present a coherent history of the industry as a whole but to illustrate, in the words of the RCHME's Royal Warrant, 'monuments and constructions ... illustrative of the contemporary culture, civilisation and conditions of life of the people in England ...'.

Thanks to the work of generations of photographers, photo-journalists and film makers, depictions of the coal industry form part of our common stock of visual images, even though mining is as remote from our own experience as it was from that of George Orwell's readers over fifty years ago. In such circumstances there is an understandable tendency for us to sentimentalise the industry and rather than confront the fact of its economic and environmental impact, the sheer scale of its endeavours and its cost in human lives and aspirations, to concentrate on the noble heroism of those involved and the poetic poignancy of the ravaged landscapes left in their wake. This book, the most recent in the Commission's continuing programme of recording the monuments of industry, eschews this approach. It confines itself to the illustration and explanation which is consistent with the making of a record for the benefit of all who are concerned to understand and to chronicle an industry which in the words of the author 'played a major role in shaping the country and transforming it into the "workshop of the world"'.

The Commissioners would like to thank all the staff involved in the project and the many people outside the Commission who assisted them, all of whom are named in the acknowledgements. They wish particularly to thank Robin Thornes for drawing together and concisely presenting a large body of material in a manner both instructive and timely.

PARK OF MONMOUTH

Acknowledgements

This survey and publication could not have been carried out without the generous assistance of a large number of individuals and organisations involved in the coalmining industry and its associated buildings.

The Royal Commission extends special thanks to Rosemary Preece of the Yorkshire Mining Museum for advice and assistance during the survey and for her valuable comments on the text. We would also like to thank the following for their assistance in the carrying out of the survey: Margaret Faull of the Yorkshire Mining Museum; Geoffrey Preece of Doncaster Metropolitan Borough Council (Recreation and Cultural Services); Penny Wilkinson of Wansbeck District Council (Leisure Department); John Gall of the North of England Open Air Museum (Beamish); the men of Frickley Colliery, West Yorkshire, and of the Royston Works of the Monckton Coke and Chemical Company; Robin Morgan, George Morgan, Garry Simmonds and Richard Hardy of Hopewell Mine; John Cornwell of Bristol Coalmining Archives Ltd; Ian Standing of the Dean Heritage Museum; Mrs P Rowe of Radstock, Midsomer Norton and District Museum; Simon Wright of L G Mouchel and Partners; Mark Higginson of Derby Industrial Museum; James Hutchinson and Jim Worgan of Chatterley Whitfield Mining Museum; Alan Davies of Salford Museum of Mining; Maureen Gee and Sarah Power of the Health and Safety Executive; V O S Jones of the Coal Industry Social Welfare Organisation; Mrs J Montgomery of British Coal Opencast (Northern Region); E M Parker, Betteshanger Colliery, British Coal; Tony Palmer of British Coal Property; David Burnhope, Coal Products Limited; Coalite Smokeless Fuels; Mr Flack and Paul Mannifield, Hay Royd Colliery, West Yorkshire; M J Sharp of New Holland Bulk Services; Ian Lavery and Danny O'Connor of the National Union of Mineworkers; Reverend Davies of All Saints' parish church, Denaby Main, South Yorkshire; Planning Departments of Bolsover District Council, Chesterfield District Council, Sheffield City Council, and Wakefield Metropolitan District Council; Nottinghamshire County Council Conservation Unit; Birtley Library, Chesterfield Local History Library, Edward Boyle Library at Leeds University, Newcastle upon Tyne Local History Library, and Nottingham Local History Library.

The work of survey and publication has involved a large number of Commission staff, working to a tight deadline. I would like to thank the following for their contribution: James Darwin, Sheila Ely and Barry Jones for fieldwork and research; Bob Skingle (RACS) for co-ordinating the photography and taking many of the photographs, together with James Davies (JOD), Peter Williams (PW), Roger Thomas (RJCT), Keith Buck (KMB), Tony Perry (ADP), Derek Kendall (DJK) and Keith Findlater (KDMF); Tony Berry for the graphics; Robin Taylor, Davina Turner and Susan Leiper for editing, and Kate Owen for managing the editing, design and production stages. Barbara Croucher prepared the index. Finally, I wish to thank the Commissioners and members of staff who gave advice and commented on the text: Professor Geoffrey Martin, Professor Gwyn Meirion-Jones, Dr Marilyn Palmer, Miss Anne Riches, Professor Charles Thomas, Dr John Bold and Keith Falconer.

ROBIN THORNES

Editorial Notes

All photographs are the copyright of the Royal Commission on the Historical Monuments of England and are held in the National Monuments Record (NMR) in Swindon. Photographs may be purchased from the NMR, quoting the reference number given in brackets at the end of the caption (eg, AA93/2552, BB91/23489).

The structure of this book is explained in the Introduction. Numbers in bold in the text refer to captions. In order to convey the magnitude of depth, etc, measurements have been given as Imperial; a conversion table is therefore given below.

CONVERSION TABLE

Imperial	Metric
1 in.	25·4 mm
1 ft	0·3048 m
1 acre	0·405 hectare
1 square mile	2·59 square kilometres
1 cubic foot	0·0283 cubic metre
1 lb	0·4536 kg
1 cwt	50·80 kg
1 ton	1·016 tonnes

Introduction

At its peak in 1913 the coal industry of Great Britain comprised around 2,600 working pits operated by 1,439 colliery undertakings. In that year the industry had a work-force of around 1,100,000, which produced 287,000,000 tons of coal, a third of which was exported. Since then the story has been one of gradual decline. At Nationalisation in January 1947 slightly less than 1,000 pits came into public ownership, these employing 718,400 men. A fall in coal consumption from the late 1950s led the National Coal Board (NCB) to embark on a programme of closures, an average of 34 pits closing every year between 1958 and 1973. In 1975 there were 241 pits still working, in 1985 the number had fallen to 133, and by 1992 there were 50 pits employing 43,800 men.

The story of the development of the coal industry is bound up closely with that of the industrialisation of the country as a whole. Coal was crucial to the Industrial Revolution, fuelling the blast furnaces of the ironworks and the engines which powered the textile mills. Coal also played an important part in the development of the country's transport infrastructure, and in particular of the canal and railway systems. Collieries sunk before the 19th century were often relatively short-lived enterprises, leaving little more than small grassed-over spoil heaps and, on occasion, isolated groups of cottages as evidence of their existence. The mines of the more recent past have, however, left a more substantial and permanent legacy, in the form of the settlements they called into being. These pit rows, model villages, and NCB estates all bear witness to an industry that has come and gone, an industry which touched the lives of many and played a major role in shaping the country and transforming it into the 'workshop of the world'.

The photographs reproduced in this book have all been taken by Commission photographers in the last five years. They record the current appearance of the wide range of structures associated with the coal industry in England – from the fragmentary remains of horse-powered 'gins' to the reinforced-concrete winding towers of the country's most modern mine. The only pit-head structure that many can readily identify is colliery headgear, the long steel legs and spoked wheels of which are often used as an icon for the industry. They also provide one of the most potent and evocative images of industrialisation, and for this reason have been used as a visual metaphor for the decline of traditional heavy industries in general. Less well known is the wide variety of other structures which a large colliery is likely to have had by the mid 20th century, these including coal screens, washing plants, power stations, ventilation fan houses, lamprooms, workshops, stores, baths, canteens and offices. An important purpose of this book is to illustrate examples of some of these types of structure, explain their purpose and place them in an historical context.

There has always been a strong temptation for those who record mining communities to dwell on their particularity, whether by stressing the heroic qualities of the miner or by emphasising the harshness of his life. By the late 19th century the popular picture of the miner was that of a noble if somewhat wild figure working hard in difficult and dangerous conditions, an image which newspaper reports of increasingly horrific mining disasters did much to foster. Similarly, the miner's wife was portrayed as a stoic and tragic figure waiting at the pit-head for news from the rescue teams working below.

Fluctuations in the coal markets from the 1870s led to periodic wage cuts and lay-offs, resulting in

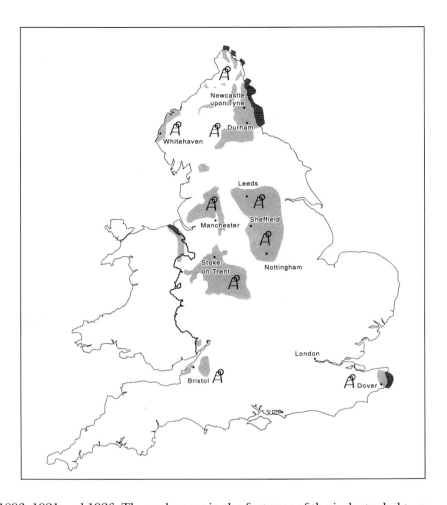

1 *Map of England showing coalmining areas.*

major industrial disputes in 1893, 1921 and 1926. These changes in the fortunes of the industry led to a change in the way it was perceived and, inevitably, to a new set of images with a new emphasis. In the 1930s, George Orwell's description of life in many mining communities in Lancashire and Yorkshire had its visual counterpart in Bill Brandt's stark and powerful photographs of miners in County Durham. In the same way that the image of headgear stood for industry in general, so the coalminer stood for the manual worker 'not only because his work is so exaggeratedly awful, but also because it is so vitally necessary and yet so remote from our experience' (Orwell 1937, 31).

World War II brought a new mood and a new set of images of the industry. These stressed the importance of coalmining to the war effort and cast the miner as hero of labour. The war artists Henry Moore, John Piper and Graham Sutherland produced images which conveyed a sense of vitality. After the war, the need to boost coal production and the heavy investment that followed Nationalisation in 1947 meant that the images of the industry continued to be positive and dynamic, a mood which did not change until the programme of pit closures began in the late 1950s.

Since the 1960s there has been a growing recognition that coal is an industry in decline, eclipsed by oil, natural gas and nuclear power. At the same time the coalminers, their culture, and their communities have come to be seen as an anachronism, survivals from a bygone age. Robert Frank, an American photographer who photographed Welsh mining communities in the decade following Nationalisation, wrote of his work:

I don't want to give an interpretation of my photographs or to give them a particular meaning, a historical meaning … I want to use these souvenirs of the past as strange objects from another age. They are partly hidden and curiously resonant, bringing information, messages which may or may not be welcome, may or may not be real. (Robert Frank 1991)

The last twenty years have seen a marked change in the way the industry has been portrayed, the images dwelling on industrial unrest and the social cost of pit closures. The 1970s saw a renaissance of social documentary photography in Britain, with photographers like John Davies, Chris Killip and Graham Smith following in the footsteps of Brandt – seeing themselves 'as artists working within that tradition which might be described as a "creative interpretation of reality"' (Kismaric 1990, 12). John Davies' photographs of post-industrial landscapes of the north of England ask the question 'what is this place now that its purpose is gone?' (Davies and Powell 1987), while Chris Killip's studies of the seacoalers of Northumberland are testaments to the lives lived in those landscapes.

The photographs taken in the course of the Commission's survey of the coal industry, a selection of which is reproduced in this book, are of structures which owe their existence to the coal industry and of the people associated with it. They are intended to give a sense of the scale of coal's impact on the built environment, by illustrating and explaining the significance of a number of individual sites. The narratives that accompany the photographs draw on the history of technology, social and economic history, and architectural history. However, this book is not a history of the coal industry, and it is for this reason that images made by previous generations have not been included. Rather, it is a record of the legacy of coal, and has in a very conscious way been made at a point in time, about a point in time.

1

The Colliery

2 *Pleasley Colliery, Pleasley, Derbyshire. Shaft sinking at Pleasley Colliery began in 1871 and the first coal was raised in 1875. The pit ceased production in 1983 when its workings were merged with nearby Shirebrook Colliery. The photograph shows the engine house of 1873 and the steel headgear above the pit's two shafts, erected in c 1900. (BB93/21394)*

3 *Free miners at Hopewell Mine, West Dean, Gloucestershire. (BB93/8474)*

Hopewell Mine, West Dean, Gloucestershire

SO 603 114

The drift mines favoured by the private miners of today are of a type once common in all areas of the country where coal outcropped on valley sides and could be won with relative ease. The Forest of Dean is one of the few areas of the country where there is an unbroken tradition of drift mining stretching back to the medieval period. It also has the distinction of being the only coalfield in Britain to be sited wholly within a royal forest. The forest once occupied most of the district which lies in the lower part of the triangle formed by the River Wye and the River Severn, but has shrunk over the centuries to an area of a little over 35 square miles. There is evidence that iron was worked in the area by the Romans, while coal has been mined since the Middle Ages. The mining of coal in the forest is still carried on by Free Miners, that is men born in the Hundred of St Briavels who have been 'bred and brought up in the mystery or craft of mining' and have worked coal for 'one year and a day in some mine within the forest' (Hart 1953, 1).

In 1788 there were 442 free miners and boys working ninety-nine mines in sixty-six companies, a company or 'vern' usually comprising four partners. By the early 19th century the industry had expanded to such an extent that disputes were becoming common. In 1831 a Commission was appointed to investigate the methods of working and examine the needs for new regulations. The outcome was the Dean Forest (Mines) Act of 1838, by which the rights of the Free Miners were codified. The Act also appointed Mining Commissioners who, in 1841, described 104 claims or 'gales' and made rules and regulations for their working. When the coal industry came into public ownership in January 1947 there was provision for the NCB to license private concerns to work coal that the NCB regarded as being uneconomic for it to mine. In 1992 there were around 138 private mines in Great Britain licensed to work coal underground, of which the great majority are in South Wales. At that time these mines employed 1,647 people and produced 1,103,000 tons of coal (Anon 1992, 18).

Hopewell drift mine [3 and 4], opened in 1976, works the Coleford High Delph seam. The seam is between 4 and 8 ft thick and produces coal of high calorific value. The mine is worked by a partnership of four: Robin Morgan, his brother George Morgan, Gary Simmonds and Richard Hardy. A bothy houses a compressor which provides compressed air for drills and jack-hammers, and an electric winch for hauling the coal tubs or 'drams' to the surface. Communication between bothy and face is by field telephone. The majority of the coal raised is sold to British Coal, who add it to imported coal in order to produce a blend which has a calorific value sufficiently high for use at power stations.

4 *Hopewell Mine: entrance to drift.*
(BB93/8469)

5 *Wearmouth Colliery, Sunderland, Tyne and Wear: view from across the River Wear showing pit waste being loaded into seagoing barge. (AA93/3720)*

Wearmouth Colliery, Sunderland, Tyne and Wear

NZ 393 579

By the 17th century the British coal industry was dominated by the collieries of Durham and Northumberland. The proximity of large reserves of coal close to the area's two principal navigable rivers, the Tyne and the Wear, had given these coalfields a major advantage over others. Coal was being shipped by sea from the north east by the middle of the 13th century, but it was not until the 16th century that exploitation began on a large scale. By 1800 the pits of the north east were responsible for around 40 per cent of English coal production (Flinn 1984, 26). Writing in the 1720s, Daniel Defoe made the following observation:

> From hence the road to Newcastle gives a view of the inexhausted store of coals and coal pits, from whence not London only, but all the south part of England is continually supplied; and whereas when we are at London, and see the prodigious fleets of ships which come constantly in with coals for this encreasing city, we are apt to wonder whence they come, and that they

do not bring the whole country away; so, on the contrary, when in this country we see the prodigious heaps, I might say mountains, of coals, which are dug up at every pit, and how many of those pits there are; we are filled with equal wonder to consider where the people should live that can consume them. (Defoe 1983, 126)

By Defoe's time the coal close to the surface had been largely worked out, with the result that colliery owners were being forced to sink deeper pits. The result was the development of larger and more capital intensive ventures, higher production being the only way to recoup the considerable investment in sinking, equipping and working the pits built to exploit deep seams of coal. By the late 18th century the average depth of pits on Tyneside was around 350 ft (Galloway 1971, 294–5).

It was in this period that trial borings were made in the east of County Durham to determine whether there was coal beneath the Magnesian Limestone. These borings did find coal, but it was not until the early 19th century that large reserves were proved to exist. Wearmouth Colliery was one of the first pits to be driven through the limestone, and the first coastal colliery in County Durham. Sinking began in 1826, but the first seam of reasonable thickness was not reached until 1834, by which time the shaft had reached a depth of 1,578 ft. Deepening of the shaft continued because of a conviction that the true seam of coal had not yet been reached. The Hutton seam was finally found in 1846 at 1,720 ft, making the pit the deepest in the country at that time (Galloway 1969, 207). To pump out the mine, a double-acting steam engine of 180 horsepower was installed in 1831. The coal was raised to the surface by a vertical steam-winding engine, built by Thomas Murray and Company in 1848 (see p 28). The cost of sinking the pit and equipping it was reputed to have been around £100,000.

The pit had its own staithes for loading coal directly into seagoing colliers, the first staithes being wooden structures on the river bank, connected to the colliery by an inclined plane. These were replaced later by a number of tall brick structures which continued in use until 1968, although one has been retained for loading pit waste into barges for disposal at sea [5]. The progressive introduction of mechanised coal cutting, loading and washing in the 1950s and 1960s resulted in a dramatic change in the ratio of coal to spoil brought to the surface at British pits. In 1920 some 10 million tons of spoil were produced, but in 1980 the amount had increased to 55 million tons. By the latter year the pits of the north east had a waste output of around 6 million tons per annum, of which 75 per cent was disposed of at sea (*Coal and the Environment*, 58).

6 *Chatterley Whitfield Colliery, Stoke on Trent, Staffordshire.* *(AA93/2552)*

Chatterley Whitfield Colliery, Stoke on Trent, North Staffordshire

SJ 885 534

In 1800 the West Midlands was second only to the north east in terms of coal production, producing 23 per cent of all coal mined in English pits (Flinn 1984, 26). The area had benefited from the growth of heavy industry in the previous century and from the accompanying development of a large canal network. By the beginning of the 19th century the coal industry of the region was dominated by the pits of Staffordshire, this county being divided into the three mining areas of South Staffordshire, North Staffordshire and Cannock Chase. From the middle of the century rising output from new and deeper sinkings in the latter two areas offset a decline in production in the former, where the small and relatively shallow pits which had characterised the area were becoming exhausted or inundated with water.

The largest pit in the history of the Staffordshire coalfield was Chatterley Whitfield Colliery in North Staffordshire [**6**], which employed over 4,000 men at its peak. The colliery has its origins in a number of small pits which existed on the site by the mid 19th century. In 1853 these pits were taken over by Hugh Henshall Williamson, whose early career had been in the pottery industry at nearby Longport. Williamson initiated a major expansion programme, central to which was the building of a railway from the pits to link up with the Biddulph Valley branch of the North Staffordshire Railway (opened for goods in 1859). By the early 1860s the pit had four shafts, then known as the 'Ragman', 'Ten Feet', 'Bellringer' and 'Laura'. The heavy debts incurred by Williamson in developing the pit led to his having to sell it in 1863 to a partnership of local gentlemen who went on to form the Whitfield Colliery Company Limited. In 1872 this company sold the pit to the Chatterley Iron Company, in whose hands it remained until the receivers were called in in 1884. A new company was then formed under the name Chatterley Whitfield Collieries Limited, from which title the pit takes its name. In the late 19th and early 20th century three new shafts were sunk: the Platt (1881–3), Winstanley (1913–14) and Hesketh (1915–17). A coal-washing plant was added in 1922 (since demolished), the pit-head baths in 1937 and the steel heapstead structures in 1954.

Coal production ceased in 1977, although the NCB continued to use the site for other purposes until 1989. In 1979 the Chatterley Whitfield Mining Museum was established, becoming the first coal-mining museum in Britain on the site of a former colliery. It was a charitable trust, drawing its income from visitor admissions, sponsorship and grants from the local authority. The museum closed in August 1993 and its future was uncertain at the time of writing.

7 *East Pit at Silverwood Colliery, Rotherham, South Yorkshire. (BB91/23489)*

Silverwood Colliery, Rotherham, South Yorkshire

SE 478 939

The coal industry of Yorkshire and the East Midlands expanded rapidly in the second half of the 19th century and by the end of the century the pits of Yorkshire and the East Midlands were, for the first time, producing more coal than those of Durham and Northumberland. This massive increase in output was made possible by a growing railway network which gave access to the expanding home and overseas markets. In that period it became apparent that there were considerable reserves of coal to the east of the existing 'exposed coalfield'. Trial borings made at South Carr, Humberside, in 1889 located the Barnsley seam at a depth of 3,186 ft. The seams of the 'concealed coalfield' were deep by the standards of the day and could only be exploited by companies with considerable capital, the pits sunk to work it being large by the standards of the time. By 1909 new pits, regarded as being 'among the most advanced in the world', had been sunk in South Yorkshire to work the Barnsley seam at Bentley, Brodsworth, Dinnington (Top Hard Coal), Frickley, Grimethorpe and Silverwood (Supple 1987, 183). The report of the Royal Commission on the Coal Industry, published in 1926, believed that 'the most important feature of the last thirty years … is the development of the great new coalfield of South Yorkshire and Nottinghamshire, where rich seams of good quality coal, but situated at considerable depths, are being actively worked' (*Report of the Royal Commission on the Coal Industry* 1926, 46).

Typical of the South Yorkshire pits of the concealed coalfield in this period is Silverwood Colliery [7], established by the Dalton Main Colliery Company at the beginning of the 20th century. The work of sinking the pit's two shafts began in April 1900 and the Barnsley seam was reached in December 1903 at a depth of 2,238 ft. By the end of the decade the pit was the largest in the Yorkshire coalfield in terms of men employed, with 2,593 men working underground and 635 on the surface. It was also believed to be the largest mine working a single seam in Great Britain (VCH 1912, 365). One of the pit's two shafts was equipped with triple-deck cages, each of which was capable of carrying a dozen 12 cwt tubs (four on each deck). Silverwood achieved its maximum output in 1929, when it is claimed 1,322,501 tons of coal were drawn from the two shafts. Three years later the pit made a production record of 7,073 tons in a day (*The Colliery Guardian* 30 October 1931). The buildings on the site included offices, workshops, wagonshops, lamp cabin, engine houses, electricity generating station, coal-washing plant and a brickworks. In addition, there were batteries of coke ovens with associated acid and tar plants (*The Colliery Guardian* 30 October 1931). The colliery company was also part owner of a number of housing schemes in the area, including the garden village of Sunnyside, a development built for the company's workers by the Industrial Housing Association (see p 87).

In the mid 1920s it was estimated that the total cost of laying out a colliery sunk to a depth of around 2,400 ft was around £1,900,000, a figure which included houses for the work-force and working capital. Sixty years before £100,000 had been regarded as sufficient to sink and equip a large colliery with an output of 250,000 tons per year (*The Iron and Coal Trades Review* 1927, 6).

Parkside Colliery, Newton le Willows, Merseyside

SJ 600 947

In the decade following Nationalisation the NCB embarked on a major programme of investment in the industry, most of which was spent on reconstructing existing pits rather than sinking new ones. One of the few new pits to be established in that period was Parkside Colliery [**8**]. An intensive programme of deep boring carried out in the Lancashire coalfield in the 1950s led to the decisions to reconstruct Astley Green, Bold, Clock Face and Sutton Manor collieries, to reopen Agecroft Colliery and to sink a new pit at Parkside. Planning permission for Parkside Colliery was granted in October 1956 and in November of the following year work began on sinking the pit's two shafts. The architect for the project was J H Bourne and the design contractors were G Wimpey and Company Limited, who were also awarded the contract for driving the 15,000 ft of underground roadways. The new pit, which it was estimated would cost £9·6m, was designed to have an output of 4,000 tons of coal per day.

By March 1959 the shafts had been completed to a depth of 2,620 ft and production began in the Crambouke seam in March 1964. The shafts were later deepened in order to work the Ince Six Foot seam at 2,920 ft and the Wigan Four Foot seam at 3,250 ft. Above each shaft was built a tall reinforced-concrete winding tower, structures which have the outward appearance of austere blocks of flats. The tower above No. 1 shaft (manriding and materials) is 190 ft high and that of No. 2 shaft (coal winding) is 204 ft high. No. 1 shaft was fitted with two four-deck cages, each capable of carrying 144 men or four 3 ton cars, and No. 2 shaft was equipped with two skips for winding coal. Both towers had two electrically driven Kœpe winding drums (see pp 55–7). Other buildings and structures on the site include a concrete framed heapstead, coal-preparation plant (with a capacity of 6,500 tons per day), power house, combined workshops and stores, powder magazine, locomotive shed and a building housing the pit-head baths, lamproom, offices, canteen and boiler house. At its peak in the mid 1970s the pit employed around 1,600 men and produced 762,000 tons of coal in a year. Production at Parkside ceased in 1992 and the pit closed in 1993.

9 *Asfordby Mine, Leicestershire.*
(BB93/28340)

Asfordby Mine, Asfordby, Leicestershire

SK 72 20

England's newest colliery is at Asfordby in Leicestershire [**9**]. In the 1970s the NCB produced a proposal for developing the Leicestershire coalfield which would have involved the construction of new mines at Asfordby, Hose and Saltby. The three proposed pits had a projected labour force of 3,800 and an expected output of 7,000,000 tons per year (Ashworth 1986, 390). The Board applied for planning permission for the three pits in 1978, but this was denied following a public enquiry. In 1983 planning approval was granted for a more limited scheme involving the building of only one colliery at Asfordby to work the Parkgate, Main Deep and Blackshale seams. Work on the building of the mine (from the 1970s the NCB has used the term 'mine' in preference to 'colliery' for new sinkings) began in the following year.

The work of sinking the mine's two shafts commenced in 1986, freezing techniques being employed to sink the shafts through water-bearing strata. The coal-winding shaft was sunk to a depth of 1,580 ft and the manriding and materials shaft to 1,670 ft. A reinforced-concrete winding tower was built above each shaft, that above the coal-winding shaft being 170 ft in height and that above the manriding shaft 154 ft. Both towers are equipped with two electrically driven Kœpe winding drums. The coal will be raised using two skips, each of 26 tons capacity. It is anticipated that the winding equipment will be able to raise up to 1,000 tons of coal per hour. The mine's coal-preparation plant is capable of processing up to 1,200 tons per hour.

Other buildings on the site include a combined workshops and stores, boiler house and a complex housing lamproom, baths, medical centre and canteen. Concern over the environmental impact of the mine has led to a considerable amount of thought being put into landscaping, including the planting of a screen of trees and shrubs around the edge of the site. The mine has been designed to have the flexibility to produce between 1,500,000 and 4,000,000 tons of coal per year, according to demand. The estimated reserves workable by the mine are around 100,000,000 tons. The coal produced will be shipped out by rail, using a loading facility designed to handle trains of up to sixty wagons. At the time of writing the mine is still under development and is not expected to start producing coal until late in 1994.

10 *Stobswood Opencast Site, Widdrington, Northumberland. (AA93/3775)*

Stobswood Opencast Site, Widdrington, Northumberland

NZ 220 940

Opencast coal extraction is a civil engineering rather than a mining operation and can only be carried out in areas where there are coal seams close to the surface. Before extraction begins the overburden must first be scraped back to reveal the coal seam, often partially worked in previous years. The topsoil is stored in mounds around the site for replacement once the coal has been worked out. Opencast mining did not begin in England until 1941, when it was embarked on as a wartime emergency measure aimed at making up for a shortfall in coal production. At the time it was viewed as a short term expedient and there was no intention that it would continue after the war was over. From the outset the work was undertaken by civil engineering firms working on behalf of government departments. In 1947 the NCB was asked if it would take over responsibility for opencast production, but in the event it did not do so until 1952. One of the problems from the Coal Board's point of view was that in the early years opencast extraction was a loss-making activity; another was that opencast mining was still being undertaken using legal powers taken by the government under wartime Defence Regulations, a legal difficulty that was not resolved until the Opencast Coal Act of 1958.

In 1947 opencast extraction produced around 10 million tons of coal, this representing 5 per cent of the total amount mined in that year. By 1991–2 fifty-eight sites were producing 16,700,000 tons of coal – 18 per cent of the total production. The average surface area of opencast sites in Britain is 523 acres and the average depth 131 ft, making them small by world standards. The working life of an opencast site averages only five years, after which it is backfilled and

11 *'Ace of Spades' working at Stobswood Opencast Site, Widdrington, Northumberland. (AA93/3773)*

landscaped. The Stobswood Opencast Site in Northumberland is larger than average, being 1,605 acres in extent and having a maximum depth of 624 ft [**10**]. Its working life expectancy of twelve years is also considerably longer than average. The site, which is worked by Crouch Mining under contract to the British Coal Corporation – in 1987 the NCB changed its name to the British Coal Corporation – also boasts Europe's biggest walking dragline excavator, the 'Ace of Spades' [**11**]. It was built at a cost of £15m, weighs 4,000 tons, has a total machine power of 15,690 horsepower, and has a bucket capacity of 1,755 cubic feet (compared with just over 81 cubic feet for the best equipment during World War II).

12 Seacoalers working on the beach at Lynemouth, Northumberland. (AA93/2658)

Seacoalers, Lynemouth, Northumberland

NZ 30 90

Sea coal has been gathered from the beaches of north-east England for centuries, having been broken away from coastal outcrops by the action of the sea, washed and then thrown up on the shore. Throughout the Middle Ages the term sea coal was commonly used to differentiate mineral coal from charcoal (both of which were referred to as coal or 'carbo' in Latin), irrespective of whether it was collected from beaches or dug from coal pits. There appear to be two separate origins of the term sea coal, one being derived from the gathering of coal on the beaches of the Northumberland coast, and the other from the fact that in the Middle Ages coal was transported to the important London market by sea. Writing in the mid 16th century, Leland commented that 'The vaynes of the se coles lye sometyme open apon clives of the se, as about Coket Island [Coquet Island, Northumberland] and other shores; and they as some will, be properly callyd se coall' (Leland 1964, 140). A contemporary of Leland, Dr Kaye, took the view that 'sea coal, or Newcastle, or smithy coal' were 'names borrowed either from the mode of its carriage, from the situation in which it is found, or from the use to which it is applied; for it is dug up in places near to New Castle, a famous city of England, and is carried thence by ships to other parts of the kingdom, and is used by smiths to soften their iron' (Galloway 1971, 109).

Today, much of the coal collected from the 'black' beaches of the north east is waste discharged by coastal pits, including Easington Colliery, County Durham, and Ellington Colliery, Northumberland. At Lynemouth the sea coal is only washed up on the beach under certain conditions of wind and tide. When the wind blows from the north east coal is washed up at the south end of the beach and when it blows from the south east the coal is washed up at the north end. The beach at Lynemouth is adjacent to British Coal's Lynemouth coal-preparation plant, which washes the coal from nearby Ellington Colliery. The gathering of the coal is carried out by seacoalers, some of whom still use horse and cart to carry the coal away from the beach [**12** and **13**]. In recent years the work and lifestyle of the seacoalers has been recorded by the photographer Chris Killip and has been the subject of the feature film, *Seacoal* (Amber Films, 1984).

13 *Horse and cart used by seacoalers at Lynemouth, Northumberland. (AA93/3708)*

2

The Architecture of the Pit

14 *Headgear, site of Western Colliery (latterly Brittains Colliery), Ripley, Derbyshire. Western Colliery was established by the Butterley Company in the late 1840s, and mined both coal and ironstone. The pit later merged with Brittains Colliery, which itself closed in 1947. (AA93/1095)*

16 *Remains of horse gin at Ram Hill Pit, Westerleigh, Avon. (BB93/8533)*

Horse gin, Ram Hill Pit, Westerleigh, Avon

ST 679 803

> Engines at pits' mouths, and lean old horses that had worn the circle of their daily labour into the ground, were alike quiet; wheels had ceased for a short space to turn; and the great wheel of earth seemed to revolve without the shocks and noises of another time.
>
> (Charles Dickens *Hard Times*)

Before the introduction of steam winding engines in the late 18th century, coal was brought to the surface by manpower, horsepower, waterpower and even by windpower. However, by the mid 18th century it was the horse-powered whim gin that was the most commonly used method of raising coal. These gins were worked by a single horse, or a team of two horses, walking round in a circle and turning a drum on to which the winding rope was wound [**15**]. *The Compleat Collier* – a description of mining in the north east, published in 1708 – was of the opinion that a pit of 360 ft in depth and with an output of 126 tons per day would require eight horses (four pairs working shifts) to wind coal (Anon 1990, 24).

The gin at Ram Hill Pit probably dates from 1830, the year in which sinking began. The pit was owned by the Coalpit Heath Colliery Company, an enterprise described in 1841 as being extensive and well regulated, with 280 hands employed. Forty or fifty of these employees were boys under thirteen years of age, some of whom were employed to draw small 1 cwt tubs from the face to the underground horseways. The raised circular platform of the horse gin, sometimes known as the 'horse race', has the remains of a stone wall around it [**16**], suggesting that originally it was roofed. The gin was later superseded or augmented by a steam engine.

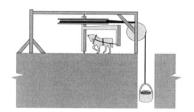

15 *Whim gin.*

Pumping engine house, Saltom, near Whitehaven, Cumbria

NX 964 174

18 Pumping engine house at site of Saltom Colliery, Whitehaven, Cumbria. (BB93/5920)

Where coal seams were higher than the ground level in nearby valleys, downward sloping adits or soughs could be driven to provide natural drainage for workings. However, when pits went deeper than was allowed by free drainage, or where such drainage was not an option, water could become a major problem. In 1610 Sir George Selby told Parliament that the growing problem of flooding meant that the pits in the vicinity of Newcastle would not last more than a generation (Galloway 1969, 53). Waterwheel-driven pumps were in use by the 16th century, and in the late 17th century Sir Thomas Liddell's Ravensworth Colliery was employing a system of three waterwheel-powered pumps to raise water up the shaft in stages (Galloway 1969, 56–7). Where utilising water power was not a possibility, horse gins were used to raise water as well as coal.

The first use of an atmospheric steam engine to work pumps at a coalmine was in 1712, in which year a Newcomen engine was installed

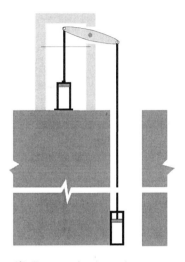

17 *Beam engine (pump).*

at a colliery near Dudley in the West Midlands [**17**]. These early engines burned large amounts of coal, but had the advantage of being relatively simple to maintain. However, they were expensive to buy and install, an engine and engine house costing in excess of £1,500. As a result they could be afforded only by the wealthier coal owners and were not installed at pits believed unlikely to repay the investment. The introduction of steam pumps meant that not only was it possible to sink shafts in areas where flooding was a problem, but also that the growing number of flooded pits could be reclaimed and brought back into production.

The pumping engine at Saltom [**18**], near Whitehaven, of which only the engine house now remains, was built to drain Sir James Lowther's Saltom Colliery, the sinking of which began in 1729. The introduction of pumping engines into the Cumbrian coalfield appears to have been the work of Carlisle Spedding, who became manager of Lowther's pits around the year 1718. Borings carried out by Spedding on the cliff side at Saltom found the Main seam at 480 ft, as a result of which the decision was taken to sink Saltom Colliery. The workings were driven out under the Irish Sea, causing Sir John Clerk to comment in 1739 that 'Sir James [Lowther's] riches, in part swim over his head, for ships pass daily above the ground where his colliers work' (Galloway 1969, 98–9). With its deep shaft and under-sea workings, the pit was regarded as one of the most remarkable mining ventures of its day. Under Spedding's management Lowther's pits prospered and by 1740 were producing 100,000 tons a year, five times their output at the beginning of the century.

Newcomen engine, Elsecar, Hoyland Nether, South Yorkshire

SK 387 999

The oldest Newcomen engine in England on its original site is the 'Great engine' at Elsecar, built in 1795 to drain Earl Fitzwilliam's New Colliery [**19**]. The engine's builder was John Bargh of Chesterfield, who was paid £69 for his work. The engine house was built by the firm of Francis Hardy and Company, who also built workshops on the site and a nearby bridge. As originally built the engine had a 42 in. cylinder, increased in 1801 to 48 in. The new cylinder was supplied by the Butterley Ironworks, a company with its own collieries and which in the 19th and 20th centuries was to play an important part in the development of the coal industry of the North Midlands. The next major rebuild of the engine was undertaken in 1835, in the course of which its wooden beam was replaced in cast iron [**20**]. The new beam, which was cast in two sections, was 24 ft long and 4 ft deep.

The engine was in continuous use until electric pumps took over from it in 1923. However, it was brought back into use for a brief period in 1928, when flooding overwhelmed and drowned the new

19 *Newcomen pumping engine house at Elsecar Main Colliery, Hoyland Nether, South Yorkshire. (AA93/1405)*

pumps that had superseded it. It was steamed again in 1931 for the summer meeting of the Newcomen Society, but was damaged in 1953 and is no longer capable of operating. The engine, which is a Scheduled Ancient Monument, was in the care of the NCB until 1988 when Barnsley Metropolitan Council bought the site.

20 *Cast-iron beam of Newcomen engine at Elsecar Main Colliery, Hoyland Nether, South Yorkshire. (AA93/1417)*

Winding engine house, North of England Open Air Museum, Beamish, County Durham

NZ 219 543

Several unsuccessful experiments at harnessing atmospheric engines to wind coal were conducted in the mid 18th century, including one at Hartley Colliery, Northumberland, in the 1760s. In the 1770s the Yorkshire engineer, John Smeaton, experimented with a hybrid solution using steam pumps to supply water to waterwheel-powered winding equipment, and in 1776 he produced a report comparing horse gins with his 'coal-engine worked by water supplied by a fire engine' (Galloway 1969, 114). 'Water coal gins' as they were known were installed at a number of pits in the north east in the last quarter of the 18th century.

The major difficulty encountered in building steam-powered winders was that of obtaining a reliable and controllable rotary motion, a problem not solved until James Watt's improvements to steam-engine design in the 1770s and 1780s. The first Boulton and

Watt winding engine was installed at Walker Colliery, Tyneside, in 1784. The widespread adoption of steam winders from the 1790s made it possible to raise larger quantities of coal from greater depths, thereby further encouraging the development of larger and deeper pits with higher outputs.

The vertical winding engine was patented by Phileas Crowther in 1800. It was the first departure from the traditional beam engine, the winding drum and flywheel being positioned directly above the cylinder [**21** and **22**]. Some of these engines were designed to be able to both wind and pump. Crowther engines, otherwise known as Durham engines, were used mainly in the north east. The positioning of the drum above the engine's single cylinder necessitated the tall engine houses, often of stone construction, that were characteristic of the coalfield in the 19th century [**23**]. Elijah Galloway in his *History and Progress of the Steam Engine*, published in 1830, commented that Crowther had constructed a number of engines of this type and that they 'were found to succeed very well' (Watkins 1955, 205). Crowther

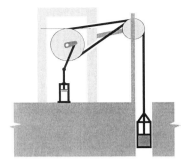

21 *Vertical winding engine.*

22 *Vertical winding engine house re-erected at the North of England Open Air Museum, Beamish, County Durham. (AA93/1357)*

23 *Vertical winding engine at the North of England Open Air Museum, Beamish, County Durham. (AA93/1359)*

engines continued to be built well into the late 19th century, the last of which may have been the one installed at Wingate colliery in 1890 (Watkins 1955, 205). The engine at Beamish was built in 1855 at the Newcastle engineering works of J and G Joicey & Company and continued in operation until 1963. In 1974 it was moved from the site of Clophill Beamish No. 2 pit to the North of England Open Air Museum in Beamish.

Winding engine house and timber headgear, Yorkshire Mining Museum, Sitlington, West Yorkshire

SE 164 254

25 *Engine house and timber headgear at Caphouse Colliery (now the Yorkshire Mining Museum), Sitlington, West Yorkshire. (AA93/1444)*

From the middle of the 19th century horizontal high-pressure winding engines were becoming increasingly common [**24**]. A well-preserved example of a small engine of this type can be seen *in situ* at the Yorkshire Mining Museum, Caphouse Colliery. The engine house at the colliery bears the date 1876 and the initials of the colliery's owner at that time, Miss Emma Lister Kaye. The building is a stone-built structure housing a twin cylinder horizontal engine supplied by Davy Brothers of Sheffield. The engine was installed when the late 18th-century No. 1 shaft was deepened to allow working of the New Hards seam. It was last used for winding coal in 1974, after which coal was brought to the surface by a newly made drift.

The timber headgear of No. 1 shaft appears to be of the same date as the engine house, making it a rare and important example of a once common type of structure. It is built predominantly of pitch

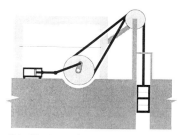

24 *Horizontal winding engine.*

pine, although the original back legs were replaced with rolled-steel ones after Nationalisation. In his *Practical Coal Mining* (1908, 82) W S Boulton commented on this type of headgear:

> In the choice of wood for pit-frames nothing will be found to give better results than pitch-pine. It is hard, strong, straight grained, and contains a large quantity of resin, which latter property renders it exceedingly suitable for use in this class of work.

Three years later the building of timber headgear at new pits was banned by the Coal Mines Act of 1911, although existing collieries were allowed to keep any already in use.

Caphouse's steam winder and timber headgear were factors in the decision for the pit becoming the home of the Yorkshire Mining Museum [25]. The museum, which is a charitable trust, opened to the public in 1988, and is now the only mining museum in England to include an underground tour of former workings.

Winding engine house, Bestwood Colliery, Bestwood, Nottinghamshire

SK 557 478

The tall buildings which housed vertical winding engines were often the most prominent and permanent structures at the collieries where they were found. It is not surprising, therefore, that more care was taken over their appearance than was given to the less substantial structures around the pit-head. This was certainly the case with the winding engine house at Bestwood Colliery [26], built around 1874 for the Bestwood Coal and Iron Company. The building is an unusually ornate structure in the then popular Italianate style. More unusual still is the fact that the building has mass concrete walls to the ground floor, an early use of this material in a colliery building. These walls are rusticated and have moulded architraves to those openings which are original. The upper floors of the building are of brick with stucco pilasters. Mass concrete was also used for the base of the large twin-cylindered vertical winding engine and for the internal iron framework that supports the winding drum above. The vertical winding engine was supplied by Worsley Mesnes Company of Wigan in 1875. This engine wound coal from a depth of 660 ft at 3 tons per wind. The shaft ceased to be used for the winding of coal following the making of a drift, but the steam winders were retained as standby units for men and materials.

Bestwood Colliery was the first pit in Britain to conduct trials with the Kœpe system of shaft winding. The system is named after Frederick Kœpe, a German mining engineer employed by Krupp.

The Kœpe winder differs from the conventional system in that both cages in the shaft are hung on a single, endless rope, this rope being passed over a deeply grooved driving pulley rather than being wound on to a drum. In early versions the driving pulley was situated in a ground-level winding house, but in later installations was often positioned at the top of a winding tower. The first Kœpe winder was installed at Hanover Colliery, Germany, in 1877. Three years later preparations were made at Bestwood Colliery for a trial of the system, although the winder did not come into use until 1883. However, the experiment was not a success, rope slippage on the drive pulley resulting in the trial being abandoned.

26 *Winding engine house, Bestwood Colliery, Nottinghamshire. (AA93/3432)*

27 *Latticework headgear, Astley Green Colliery, Tyldesley, Greater Manchester. (AA93/1691)*

Lattice steel headgear, Astley Green Colliery, Tyldesley, Greater Manchester

SD 705 999

Improvements in the technology of iron and steel making in the second half of the 19th century made those materials both more affordable and available. By the 1870s iron and, increasingly, steel headgears were becoming more common, although as late as the 1880s C M Percy's *Mechanical Engineering of Collieries* was dismissive of the use of the new materials, commenting that 'except at larger collieries, timber is likely to be as much the rule in the future as in the past'. Percy did acknowledge that 'Iron makes a firmer structure, and avoids danger from fire', but went on to argue that 'Timber is very much cheaper, and will do well enough for a term of thirty years, which is not a bad average life for a colliery' (Percy 1888, 86).

28 *Latticework headgear, Washington 'F' Pit Museum, Washington, Tyne and Wear. The headgear of the former Washington 'F' Pit appears to have been built in the first decade of the 20th century, but is of a type that dates back to the third quarter of the previous century. (AA93/4661)*

29 *Tandem headgear, Snibston Discovery Park (formerly Snibston Colliery), Coalville, Leicestershire. Tandem headgear serving a pair of shafts was once a distinctive feature of the Leicestershire and South Derbyshire Coalfield, but that at Snibston is now the only example to survive in situ. This headgear, which is of rolled-steel construction, was built in 1946 to replace a 19th-century timber structure of the same type. (BB93/28225)*

However, even before the ban on building in timber was imposed by the Act of 1911, a growing number of pits were being equipped with steel headgear. This was particularly true at new deep sinkings where it was the intention to employ the high-speed winding of large, multi-decked cages. The two principal types of steel headgear were those built of latticework and those of rolled steel, the former being the more common until after World War I. Both types had advantages, latticework headgear having the least wind resistance while rolled-steel required less maintenance.

The latticework headgear of the former Astley Green Colliery was supplied by Head Wrightson of Thornaby and Stockton (Cleveland) in 1912 and is a good example of the steel headgear of the period [27 and 28]. The pit was established by the Pilkington Coal Company (a branch of the Clifton and Kersley Coal Company), work on sinking beginning in 1908. The shaft was completed in 1912, by which time it had reached a depth of 2,670 ft. The winding engine for the shaft was supplied by Yates and Thom of Blackburn in the same year.

The colliery was closed in 1970 and most of the buildings on the site demolished. The headgear and engine house of No. 1 shaft were saved by Lancashire County Council, who owned the site until its transfer to Greater Manchester County Council in 1974. The site is currently owned by Wigan Metropolitan Borough Council and is occupied by the Red Rose Steam Society, which is engaged in restoring the winding engine which survives *in situ*.

31 *Ventilation furnace and chimney, Golden Valley, Bitton, Avon. (BB93/8554)*

Ventilation furnace and chimney, Golden Valley, Bitton, Avon

ST 690 710

As pits were sunk deeper and underground workings became more extensive, the problem of ventilation became increasingly serious. The need for good ventilation was particularly important when explosive 'firedamp' (carburetted hydrogen or marsh gas), suffocating 'black-damp' (a mixture of nitrogen and carbon dioxide), or poisonous 'stinkdamp' (hydrogen sulphide) were present in dangerous quantities. Writing in the mid 16th century, Dr Kaye noted that in the north of England there were 'certain coal pits, the unwholesome vapour where-of is so pernicious to the hired labourers, that it would immediately destroy them, if they did not get out of the way as soon as the flame of their lamps becomes blue, and is consumed' (Galloway 1971, 109).

One solution to the problem was to sink separate air shafts in order to produce a natural circulation of air through the workings, the 'downcast shaft' drawing the air in and the 'upcast shaft' expelling it [**30**]. At some pits the natural flow of air was augmented by siting furnaces at either the top or bottom of the ventilation shaft. The suction effect of the furnace drew air into the pit, a system of doors ensuring that the fresh air passed through all the workings, rather than simply taking the shortest route between the two shafts. The alternative to having two separate shafts was to have a single shaft with a partition down the middle to separate the upward and downward flowing currents of air.

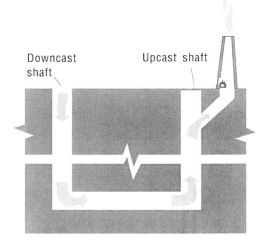

Downcast shaft

Upcast shaft

30 *Ventilation furnace.*

The ventilation furnace and chimney at Bitton [**31**] was built in the 1840s and is one of the last remaining examples of this type of structure in the country. It was erected to ventilate New and Old Pit, two of a group of coalmines in the Golden Valley to the east of Bristol. Beneath the furnace is a drift sloping down to a nearby mineshaft. The pits of the Golden Valley needed a system of ventilation not because of gases – a problem from which they were free – but because they were working a seam 18 in. thick at a depth of 1,920 ft. In most parts of the country it would not have been economic to work such a thin seam at such depth. However, the almost anthracite-like qualities of the Parrot Seam meant that the coal had a ready market and commanded high prices. The Golden Valley pits closed in 1898 and the chimney and furnace were restored to working order by the Bristol Industrial Archaeology Society almost a century later in 1984 (Cornwell 1991).

33 *Guibal fan house, Duke Pit,*
Whitehaven, Cumbria.
(AA93/1436)

Guibal fan house, Whitehaven, Cumbria

NX 969 182

The first half of the 19th century saw a growing interest in finding ways of ventilating mines by mechanical means. In 1807 John Buddle installed a steam-driven air pump at Hebburn Colliery on Tyneside. However, Buddle's pump was capable of exhausting only around 6,000 cubic feet of air per minute from the workings below, whereas the average air circulation in a mid 19th-century pit was calculated to be between 30,000 and 50,000 cubic feet per minute (Hinsley 1972, 31). More successful was William Price Struve's air pump, the first of which was installed at Eaglebush Colliery, South Wales, in 1849, and found capable of circulating 56,000 cubic feet of air per minute. The alternative to the air pump was the centrifugal fan, one of the earliest of which was installed at Hemingfield Pit in South Yorkshire by Benjamin Biram around the year 1836. In the following thirty years a number of other centrifugal fans [**32**] were developed, included among which are those designed by Brunton (1849), Nasmyth (1854), Guibal (1859), Lemielle (1860), Schiele (1863) and Waddle (1864).

Of all the fans developed in this period, the most popular was that designed by Theopile Guibal of Mons, Belgium. Guibal was the first to recognise that the blades of the fan needed to be encased if they were to work to maximum effect. His fans, which varied in diameter from 32 to 50 ft, usually comprised eight or ten wooden paddles attached to cast-iron arms. The power was provided by a steam engine alongside the fan casing. The first Guibal fan in England was erected at Tursdale Colliery, County Durham, in 1859. Three years later Guibal took out patents in Britain for his fan and in 1863 a modified

34 *Candlestick chimney, Whitehaven, Cumbria. (AA93/1447)*

and improved version was installed at Elswick Colliery on Tyneside. The Guibal fan proved popular and by 1876 there were almost 200 at work in British coalfields, the largest being capable of exhausting 200,000 cubic feet of air per minute.

The Guibal fan house at the former Duke Pit, Whitehaven [**33**], was built around 1862 and originally housed a fan of 36 ft in diameter (Falconer 1980, 74). The stone-built structure is in the form of a church, its outlet or évasée having the outward appearance of a crenellated tower. Other colliery structures around the harbour at Whitehaven were given similar treatment, these including a candlestick chimney built in 1850 by Sidney Smirke (Falconer 1980, 74) [**34**].

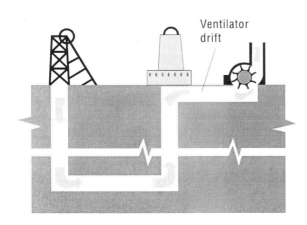

Ventilator drift

32 *Ventilation fan.*

35 *Cappel fan house, Woodhorn Colliery Museum, Ashington, Northumberland. (AA93/1012)*

Cappel fan house, Woodhorn Colliery Museum, Ashington, Northumberland

NZ 289 881

Towards the end of the 19th century smaller, faster-running centrifugal fans tended to replace the larger and slower fans such as the Guibal and Waddle. One such was the Cappel fan, invented by the Reverend George Marie Cappel. Cappel first became interested in mechanical fans as a means of drying farm produce, but in the early 1880s turned his attention to their application in the ventilation of mines. The first Cappel fan to be used in the industry was installed at Westwood Colliery near Sheffield and underwent tests at that pit in 1886. The Cappel fan operated by drawing air into a cylindrical chamber, and then forcing it out through holes in the walls of the chamber at high centrifugal force.

The fan at Woodhorn Colliery, installed in 1900, originally was powered by a horizontal steam engine supplied by Robey and Company of Lincoln, a firm with a reputation for building horizontal engines that were both fast and reliable. The engine at Woodhorn was capable of turning the fan's 16 ft diameter chamber at 160 revolutions per minute, at which speed the fan extracted 120,000 cubic feet of air per minute from the workings below. (Following the replacement of the steam engine with a 300 horsepower electric motor the speed of rotation was increased to 225 revolutions, making the fan capable of evacuating 180,000 cubic feet per minute.) The building which houses the fan is of yellow Ashington brick. The engine house [35] has gauged arches to doors and windows and a cogged and stepped eaves cornice. The curved outline of the fan casing and évasée is visible behind the engine house. Woodhorn Colliery closed in 1981 and is now a mining museum run by Wansbeck Borough Council.

Blacksmith's shop and stables, Woodhorn Colliery Museum, Ashington, Northumberland

NZ 289 881

The increasing size of underground workings at larger pits by the 18th century meant that coal won at the face had to be transported for greater distances underground. One solution to the problem was to build wooden railway systems underground and use horses to draw the wagons. Horses were working underground in larger pits in the north east of England by the middle of the century. In 1787 a total of 183 horses was employed at Lowther's Howgill and Scalegill collieries in Cumbria – seventy-two of which were used for underground haulage (Flinn 1984, 106). At pits where the only access to the workings was down shafts (rather than drifts) the horses were stabled underground, only coming to the surface for annual holidays and during strikes and lockouts.

The first legislation to protect horses working underground was the Coal Mines Regulation Act of 1887, which allowed mines inspectors to investigate the treatment and working conditions of ponies. In 1911 new legislation was introduced following investigations by animal welfare groups such as the National Equine Defence League (Yorkshire Mining Museum 1992). This legislation, popularly known as the 'Pit Ponies Charter', laid down that daily records be kept on horses working underground, that horses be at least four years old before entering the pit and that for every fifteen horses there should be one horseman to look after them. It also prohibited the use of blind horses. The conditions in which horses worked underground were regulated further by legislation enacted in 1949 and 1956 (Yorkshire Mining Museum 1992).

In 1913 some 70,000 ponies were working underground in British pits, the number falling to 21,000 by Nationalisation in 1947. Between the wars Britain lagged behind its major competitors in the area of underground haulage. British pits continued to rely on horses and rope haulage in a period when underground locomotives and conveyor belts were being widely adopted elsewhere. In 1929 only 14 per cent of output was being mechanically conveyed underground, rising to 51 per cent by 1937. In 1992 there were still twenty-four pit ponies working underground at Ellington Colliery, Northumberland, salvaging equipment from abandoned faces.

38 *Stained glass medallion in NUM offices, Barnsley, South Yorkshire. (AA93/3446)*

37 *Surface stabling for pit ponies, Woodhorn Colliery Museum, Ashington, Northumberland. (AA93/1010)*

39 *Coal screens, Hopewell Mine, West Dean, Gloucestershire. (BB93/8484)*

Coal screens, Hopewell Mine, Gloucestershire

SO 603 114

Screening or sieving coals appears to have been first adopted by coal owners of the north east around 1760. The purpose of screening was to separate the large coal demanded by the all important London market from the small coal [**39**]. John Holland, writing in 1835, described a typical screen of that time:

> Connected with every pit in the neighbourhood of Newcastle is a contrivance for screening the coals. In most cases it consists of a platform sloping at an angle of about 45 deg. from the raised bank about the pit toward the ground. At intervals are inserted grates 12 or 15 feet in length and about four feet wide, having the spaces between the bars more or less considerable according to the size of coal required to pass through. (Holland 1841, 209)

The coal tubs were brought out of the pit and on to an elevated platform or bank from which their contents were tipped down on to the screens. These raised structures, which often took the form of covered sheds at the pit-head, were and still are known as heapsteads.

As pits grew larger and more productive, it became essential to find ways of speeding up the screening process without breaking the larger and more saleable pieces of coal. The result was the introduction of mechanical 'jigging screens' which provided enough movement to speed up the process, but which did not damage the coal. The jigging screen, which came into general use in the 1880s, was an iron plate pierced with holes; it was shaken in order to sieve the coal passing over it. A screening plant would have a number of screen plates with different sized holes in them to separate the various sizes of coals. After screening the coal still had to be examined by coal-pickers, whose job was to remove stone, shale, pyrites ('brass') and

40 Mrs Sheila Truman at work in the family's coalyard near Awre in the Forest of Dean, Gloucestershire. This study recalls the 19th-century photographs taken of the 'pit-brow lassies' of Lancashire. (BB93/8516)

41 *Stained glass medallion depicting coal being tipped into railway trucks, NUM offices, Barnsley, South Yorkshire. (AA93/3451)*

other unwanted matter from it. From the late 19th century it was becoming common for the coal from the screens to be carried on a conveyor belt past the lines of pickers.

In most coalfields the work of coal preparation was undertaken by old men and boys and in others by women [**40**]. By the 1880s approximately 2,597 women were working on the surface in English pits, 63 per cent of whom were employed in the Lancashire coalfield. Margaret Park, Mayoress of Wigan, wrote in 1886 of her surprise at encountering 'pit brow lassies': 'When I first came into this district I was shocked at the spectacle of the trousered women looking and working like men: but I soon recovered from the shock. I found the women healthy and honestly employed' (John 1984, 31). In that year an attempt was made to stop women working on the surface at collieries. The women fought back, organised a campaign against the proposed legislation, and persuaded the Home Secretary that 'there was nothing to justify the House in interfering with an honest and healthy industry, which has been adopted by the deliberate preference of those engaged in it' (John 1984, 37). In 1918 there were more than 11,300 women working in the industry. By 1950 the number had declined to less than 1,000, most of whom were working in Lancashire. The last pit woman retired, in Cumbria, in 1972.

Heapstead and headgear, Bentley Colliery, Bentley with Arksey, South Yorkshire

SE 570 074

42 *Heapstead and headgear of No. 1 shaft, Bentley Colliery, Bentley with Arksey, South Yorkshire. (BB91/23450)*

From the early 20th century an increasing number of heapsteads and other structures around the pit-head were built using rolled steel and reinforced concrete. An early example of a reinforced-concrete structure is No. 1 heapstead at Bentley Colliery, South Yorkshire [**42**], built in 1912. The colliery's owners, Barber Walker and Co, gave the contract for building the heapsteads, headgear and a number of other structures to the French engineer, L G Mouchel. The system used by Mouchel was one patented by another French engineer, François Hennebique, in 1892. In 1921 it was estimated that there were over 3,000 Mouchel-Hennebique structures in Great Britain and Ireland, including warehouses, factories, chimneys, bridges, docks, slipways and even ships.

Bentley Colliery had two heapsteads, only one of which survives intact. Both had elegant classical elevations, giving them the appearance of being built of rendered brick rather than reinforced concrete.

43 *Reinforced-concrete headgear of Nos 1 and 2 shafts, Bentley Colliery, Bentley with Arksey, South Yorkshire. (BB93/23782)*

The upper floors have tall steel-framed windows with semicircular arched heads, while the open areas below have rows of arched openings, each with a keystone.

No. 1 (downcast) shaft was surmounted by headgear of rolled-steel construction. More unusual, though, was the headgear above No. 2 (upcast) shaft which, like the heapstead on which it stands, is of reinforced concrete. The choice of reinforced concrete rather than steel for the headgear of No. 2 shaft is likely to have been made because it was the upcast shaft, and there was, therefore, a need to enclose the top of the shaft in order to prevent air from the surface entering it and inhibiting the upward flow of air from the workings below.

The greater part of No. 2 heapstead was demolished a number of years ago, but No. 1 heapstead and its steel headgear and the concrete headgear of No. 2 shaft still survive [**43**].

Coal-washing plant, former Allerton Bywater Colliery, West Yorkshire

SE 425 277

44 *Coal-washing plant, former Allerton Bywater Colliery, West Yorkshire. (BB91/23465)*

From the late 19th century manual coal-picking began slowly to be replaced by mechanised coal-washing plants [44]. These washers worked on the principle that objects of differing specific gravity fall through water at different speeds, for a given size. By ensuring that the water through which the objects were falling was flowing at a predetermined speed it was possible to separate coal and shale of different sizes and to wash it to remove dirt. One of the most popular coal-washers of the period was the German designed Baum washer, a system which had the advantage that the coal could be washed without first being graded. In Britain the patent rights to the Baum washer were acquired by the firm of Simon-Carves, who built the first 'British Baum' washer around 1903. By 1914 the firm had erected plants at around fifty pits. In that year only 15 per cent of coal produced by British pits was washed. Coal-washing plants became increasingly common sights at collieries in the inter-war years, but in

1945 the Reid Report noted that less than half of Britain's coal output was mechanically cleaned. The NCB's *Plan for Coal* (1950) recognised this deficiency and declared the Board's intention to embark on 'a heavy programme' of building coal-preparation plants.

Among the first fifty Baum coal-washing plants to be built by Simon-Carves was an installation at Allerton Bywater Colliery. The washery was powered by a Westinghouse electric motor of a total of 190 horsepower and was capable of washing 75 tons per hour. The washing plant was later enlarged and by 1954 there were two Baum washers, one of 120 and one of 80 tons capacity. Originally the slurry, comprising fine coal and dirt, was pumped into settling ponds and eventually reclaimed for use in the boilers. However, in 1949–50 a plant was erected to clean the slurry and recover the fine coal. Allerton Bywater Colliery ceased producing coal in 1992.

Generating station, Philadelphia, Houghton le Spring, Tyne and Wear

NZ 335 520

45 *Generating station, Philadelphia, Houghton le Spring, Tyne and Wear. (AA93/1387)*

The first electricity generating plant at an English pit was installed at Trafalgar Colliery in the Forest of Dean in 1882. By the 1890s electricity was in use in a number of pits, although the new technology was given only a cautious welcome by the coal industry. The views of many were summed up in Percy's *Mechanical Engineering of Collieries* (1886), which declared:

> The advocates of electricity as a means of transmitting power cannot expect mining engineers to accept their mere opinion that it will be found efficient. All new contrivances must rest upon their merits, they must be weighed in the balance of work done, and will be adopted just in proportion as actual proof is given that they are better, or, at any rate, as good in efficiency and economy as the other systems so long in actual use. Up to the present time electricity has been useful at some collieries for purposes of light, and for signalling, but scarcely as a transmitter of power.

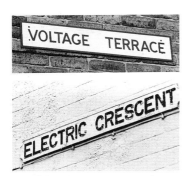

46, 47 *Street signs in vicinity of generating station at Philadelphia, Houghton le Spring, Tyne and Wear. (AA93/1388 and AA93/1389)*

In 1902 there were 145 electrically powered coal cutters in British pits, rising to around 2,000 in 1912. By 1913 only some 50 per cent of mines were using electricity, including 272 in the north east of England. Nonetheless, the use of electricity was regarded as significant enough for an Electrical Inspector of Mines to be appointed in 1909 to monitor the application of electric power in the industry.

The generating station at Philadelphia was built around 1906 for the Durham Collieries Power Company (Linsley 1976) [**45**]. It was sited alongside a temporary station opened by the Sunderland District Tramways Company in the previous year. The major user of the new station was the Lambton Coal Company, which had a number of collieries in the area. By 1911 the station had been incorporated into the Newcastle Electricity Supply Company's system, which by then had around fifteen generating stations and was distributing power as far north as Ashington in Northumberland and as far south as Teesside. The surviving generating hall is a large building of yellow brick with red brick dressing. Inside, a cast-iron balcony runs along the south side of the hall, with cast-iron staircases to ground floor and basement. The novelty of the generating station when first built is evident from the fact that one street in the vicinity is named 'Voltage Terrace' and another 'Electric Crescent' [**46** and **47**].

No. 1 Winding House, Ledston Luck Colliery, Ledston, West Yorkshire

SE 429 307

48 *No. 1 Winding House, Ledston Luck Colliery, Ledston, West Yorkshire. (BB91/21088)*

Electric winding engines were introduced into English pits at the beginning of the 20th century, one of the first being installed at Harton Colliery, Tyne and Wear, in 1908. The years leading up to the outbreak of World War I saw only a gradual increase in the number of electric winders, many companies continuing to prefer steam engines. Among the first electrically driven winders to be built in this period was that installed at Ledston Luck Colliery, West Yorkshire, in 1911 [**48**]. In 1912 the *Victoria County History* noted that:

> At a new sinking at Ledston by the Micklefield Coal and Lime Company electricity from the power company is to be solely employed for lighting, underground haulage, pumping, driving of surface machinery, and for shaft winding. This will be the first electric winding plant to be put into operation in Yorkshire at a main winding shaft. (VCH 1912, 363)

The electricity used at the pit was supplied by the Yorkshire Electric

Power Company, whose power stations were at Thornhill (near Dewsbury) and Barugh (near Barnsley) (*The Colliery Guardian* 15 May 1914). By the time the winders came into use the power company was supplying electricity to thirteen Yorkshire pits, some of which also had generating plants of their own.

The pit's No. 1 (downcast) shaft was equipped to wind 1,500 tons of coal in a working day of ten hours' duration. The winding equipment comprised two DC motors driving a winding drum of 11 ft diameter. The electric winding equipment was built by the Lahmeyer Electrical Company (by 1914 the AEG Electrical Company). The winding house of No. 1 shaft is an unusually ornate building in brick with sandstone ashlar. The building (and the gates nearby) shows the influence of the Arts and Crafts movement, while the massive mullion and transom windows and parapet combine to give it the appearance of an Elizabethan 'prodigy' house.

49 *Kœpe overhead winding tower at Murton Colliery, Murton, County Durham. (BB92/9086)*

Winding tower, Murton Colliery, Murton, County Durham

NZ 403 474

The first tower mounted Kœpe winder in Great Britain was installed at Plenmeller Colliery, Northumberland, in 1914. Its success led the South Hetton Coal Company to commission an overhead Kœpe winder for West Pit at Murton Colliery, County Durham, the installation coming into use in December 1923 [49]. The principal contractor for the project was Metropolitan Vickers Electrical Company, Manchester. Much of the mechanical work was subcontracted by Vickers to Cowans, Sheldon and Company, Carlisle, and the building of the tower to the British Reinforced Concrete Company, Manchester. The tower, which was built over the colliery's upcast shaft, is a fireproof structure with reinforced-concrete frame, floors and roof. The walls are of brick and the stair and fire escape of steel.

One of the advantages of housing the shaft's 420 horsepower winding motor at the top of a tower rather than in a ground-level engine house was that it significantly reduced the wear of the winding rope by eliminating the harmful wave motion characteristic of the latter arrangement [50]. When lowering men and materials, the winder's electric motor acted as an alternator, thereby generating current for the supply system and providing additional braking. The cages at Murton had four decks, each deck holding two 10 cwt capacity tubs. Despite its advantages and its popularity on the continent the Kœpe winder was not widely adopted in England until after World War II, critics regarding the system as less safe than having a separate rope for each cage.

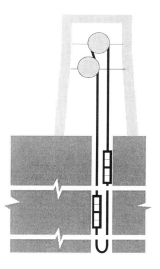

50 *Kœpe winding tower.*

51 Winding tower, Maltby Colliery, Maltby, South Yorkshire. (AA93/241)

Winding tower, Maltby Colliery, Maltby, South Yorkshire

SK 552 925

Two years after Nationalisation little more than 50 per cent of the power used at collieries was electrical, the remainder being largely produced by steam (Ashworth 1986, 99). The Board's policy of modernising pits included an extension of the use of electricity, concentrating its resources on those pits undergoing reconstruction or which had long life expectancies. As part of this programme electrically powered overhead winding towers were built at a number of pits, the first being installed at Bradford Colliery, Greater Manchester in 1951 (since demolished). These towers used the Kœpe system, the advantages of which were stressed by the Reid Report on the industry in 1945.

One of the most recent Kœpe winding towers to be built in England is that at Maltby Colliery, South Yorkshire [51]. In 1981 the NCB announced its intention of sinking a new shaft at Maltby to work the Parkgate seam. When launched this scheme was the biggest investment by the Board at a fully operational pit. A new shaft (No. 3) was sunk to a depth of approximately 3,251 ft in order to raise coal from the Parkgate and Swallow Wood seams. The shaft is 26 ft 6 in. in diameter and is equipped with four 25 ton skips in two pairs. The shaft's concrete winding tower houses two 4,000 horsepower electric motors, one for each pair of skips. The installation was the first of its kind in Britain and possibly in Europe. The first coal was wound up No. 3 shaft in 1988 at which time the target was to raise 50,000 tons per week (around 2,500,000 tons a year). The reinforced-concrete winding tower at Maltby Colliery is one of a number built at British Coal pits in the 1980s, others being erected at Harworth Colliery, Nottinghamshire, and Asfordby Colliery, Leicestershire.

52 Lamproom, Bentley Colliery, Bentley with Arksey, South Yorkshire. (BB91/23444)

Lamproom, Bentley Colliery, Bentley with Arksey, South Yorkshire

SE 570 074

The early 19th century saw a number of experiments with designs for flame lamps that could burn in the presence of explosive gas (firedamp) without igniting it. The earliest of these were carried out between 1811 and 1813 by Dr Clanny of Sunderland. The need for such lamps was highlighted by the explosion at Felling Colliery, Tyne and Wear, in May 1812, which claimed the lives of ninety-two men and boys. Accounts of the disaster were written by a local clergyman, the Reverend J Hodgson, who drew public attention to this important aspect of mines safety. Within months a society was established for preventing accidents in coalmines, the first meeting of which was held at Sunderland in October 1813. The Sunderland Society approached Sir Humphry Davy, who agreed to look into the possibility of developing a safety lamp. In October 1815 Davy wrote to Hodgson that his experiments with enclosed lamps had proved:

Atmospherical air, when rendered impure by the combustion of a candle, but in which the candle will still burn, will not explode the gas from the mines; and when a lamp or candle is made to burn in a close vessel, having apertures only above and below, an explosive mixture of gas admitted merely enlarges the light and gradually extinguishes it without explosion. (Galloway 1969, 167)

Meanwhile, George Stephenson was also experimenting with safety lamps, tests on which were carried out at Killingworth Colliery (where Stephenson was engineer) in November 1815. In the following year Davy's lamps were tried out in a number of collieries and found to be successful. In September 1817 the coal owners of the north east acknowledged Davy's contribution to mines safety by presenting him with a service of plate worth £2,500, while Stephenson was voted 100 guineas for his efforts. However, there were reservations about using these early safety lamps, and accidents in which they apparently played a part resulted in continuing research and the development of improved designs. Today, flame safety lamps are still issued to mine officials for the purpose of gas detection.

The introduction of safety lamps resulted in the appearance of a new type of building at pits: the lamproom, or lamp cabin as it was otherwise known. These buildings typically comprise a room for the storage of the lamps and another for their maintenance. The lamproom at Bentley Colliery was built in 1912 and is a fireproof structure, the floors and internal frame being of reinforced concrete and the outer walls of brick [52]. The building was designed by the colliery company's engineer, Robert Clive, and built by the firm of L G Mouchel.

53 *Stained glass medallion depicting safety lamp, NUM offices, Barnsley, South Yorkshire. (AA93/2664)*

54 *Lamproom, Frickley Colliery, South Elmsall, West Yorkshire: miners coming off shift. (AA93/3751)*

Lamproom, Frickley Colliery, South Elmsall, West Yorkshire

SE 09 46

A portable electrically powered lamp for use in mines was patented as early as 1859, but the use of electric lamps did not become widespread until after World War I. In 1914 there were 679,572 flame lamps in use underground at British pits and only 75,707 electric lamps. The relative merits of flame and electric lamps were still being debated in that year, the opinion of William Maurice, an advocate of the latter, being that:

> ... there is no invention which is likely to have so beneficial an influence on labour conditions, no invention which is so clearly destined to improve

the miner's lot, no invention that gives such bright promise of reducing the risks of underground work to anything like the same extent, as the portable electric lamp.

(Transactions of the Institution of Mining Engineers XLVIII, 286)

Eleven years later, in 1925, flame lamps still outnumbered electric lamps 526,916 to 365,823, although by then electric lamps were already in the majority in Yorkshire and the East Midlands (*The Iron and Coal Trades Review* 1927, 45). The hand-held types of electric lamp were the most common at this time, the cap lamp having only recently come into production. The electric cap lamp only gradually superseded the hand lamp, there being resistance to its use in many areas. Its advantage was that it provided illumination where it was needed, thereby combating miners' nystagmus, an eye condition caused by working in low light.

From the 1930s it became common to locate lamprooms in pit-head baths buildings [**54**], the miners removing their lamps and putting them on the charging racks before proceeding to the showers. This is the case at Frickley Colliery, West Yorkshire, where the baths block was built in 1938 by the Miners' Welfare Fund (see p 64).

55 *Pit-head baths, Gibfield Colliery, Coal Pit Lane, Atherton, Greater Manchester. (AA93/1684)*

Pit-head baths, Gibfield Colliery, Atherton, Greater Manchester

SD 665 033

Nowhere was the difference in attitude to miners' welfare between Britain and her continental neighbours more evident than over the provision of pit-head baths. By the beginning of the 20th century baths were already common in Belgium, France and Germany, and by 1914 legislation had made their provision and often their use compulsory in all three countries. However, in England the situation was quite different, with very few baths having been provided by the end of the 19th century. In 1907 the Royal Commission enquiring into the mining industry reported in favour of the establishment of pit-head baths, but the Coal Mines Act of 1911 stopped short of making their provision compulsory. Instead, the legislation stated that baths should be provided if a majority of the work-force voted in favour of having them and agreed to pay half the cost of their maintenance. In the event only six baths were provided by colliery companies in the decade following the passing of the Act.

One of the few pit-head baths to be erected in that period – and possibly the only ones to survive – are those built in 1913 by the Fletcher Burrows Colliery Company for the workers at their Gibfield Colliery [**55**]. Before building these baths, the company sent a delegation to visit pit-head baths installations in Belgium and France. One of the delegates, Clement Fletcher, commented that 'the continental baths are as a rule fitted up with quite unnecessary luxuriousness' (Preece 1988, 33). Nonetheless, the baths built by Fletcher Burrows were, in many important respects, based closely on continental practice, having a large central dressing hall and showers in side aisles. The dressing hall, which was open to the roof, was lit by windows in both the gables and clerestory. The baths also followed continental practice in providing an arrangement by which the miners could attach their pit clothes to cords and then haul them up to the ceiling, where they could be dried in the warm air rising from below, a system still in use in some Eastern European coalmines. The cords could be secured in place by padlocks, thereby preventing theft of the clothes. At Gibfield the rows of pulley wheels over which the cords passed are still *in situ*. As built, the baths had forty shower cubicles in two rows, each cubicle being 6 ft by 3 ft 4 in. and having a lockable wooden door for privacy.

The baths were an immediate success, proving popular with the company's work-force. Interviewed one month after their opening, a miner told the *Manchester Evening News*:

> It seems to make our time above ground two hours longer … Before we could wash here, we went home at 3 o'clock in our dirty clothes, and perhaps did not change them and wash ourselves until five. Now, when we come up we are dressed for the evening a quarter of an hour after leaving the cage. (Preece 1988, 36)

They were widely publicised – including a feature article in *The Colliery Guardian* (2 January 1914) – and served as the model for a number of baths built in the following decade. Gibfield Colliery closed in 1963, since when the bath house has been used for a number of purposes, the most recent of which is as a car repair workshop.

56 *Shower cubicles, Silverwood Colliery, South Yorkshire. (BB91/23488)*

57 *Pit-head baths, former Elemore Colliery, Hetton, Tyne and Wear. (AA93/1086)*

Pit-head baths, Elemore Colliery, Hetton, Tyne and Wear

NZ 355 454

One of the most important results of the Mining Industry Act of 1920 was the establishment of the British Miners' Welfare Fund, the income of which was to be used for 'purposes connected with the social well-being, recreation and conditions of living of workers in or about coal mines'. At the outset the Fund gained its income from a levy of 1*d* on every ton of coal mined (reduced to ½*d* per ton in 1934), which money was to provide grants in the general areas of health, recreation and education. In 1926 a separate fund was established for the purpose of building pit-head baths. The Baths Fund was paid for by the introduction of a levy of 1*s* on every £1 of royalties paid on the production of coal.

The intention in establishing the Baths Fund was that every pit would have baths by 1945. In 1927 the Miners' Welfare Committee followed in the footsteps of Fletcher Burrows by sending a deputation to examine recently built pit-head baths in Belgium, France and Germany. Among the buildings provided by the Fund in the 1930s were a number of impressive examples of the architecture of the Modern Movement, including designs by F G Frizzell, C G Kemp and J W M Dudding. Commenting ten years after the establishment of the Baths Fund, the annual report of the Welfare Commission for 1936 claimed that the pit-head baths built in that period were 'notable examples of the change of outlook in industrial planning' (Miners Welfare Fund Annual Report 1936, 31).

The bath buildings of the 1930s differed from those of the Gibfield type in that they often provided a number of other facilities, including boot cleaning and greasing rooms, lamproom, canteens and medical centres. The Miners' Welfare Committee favoured lockers over the cord and pulley system employed at baths of the Gibfield type. It was usual for the buildings to have two locker rooms: a dirty locker room for pit clothes and a clean locker room for day clothes, the former heated to dry wet pit clothes. The Fund gave thought to all aspects of the design of their baths, including such details as the

58 *Easington pit-head baths, County Durham. Designed by F G Frizzell and built by the Miners' Welfare Committee for the Easington Coal Company. When work began in 1935 the building was to be of two storeys, but a third storey was added prior to its opening in February 1937. (AA93/1335)*

form of the shower cubicles. In 1932 after much discussion it came down in favour of the open fronted shower stall [56], on the grounds that this type was easiest to clean and allowed miners to scrub each other's backs. At this time miners were required to wear 'modesty slips' while bathing, failure to do so resulting in a fine. By Nationalisation only around one-third of pits had baths, although these did account for 60 per cent of the work-force. Small pits and ones with limited lives tended not to have baths, their owners believing that their provision was not worthwhile. The Fund recognised that small pits did pose a problem, producing a number of designs for bath houses to serve as few as twenty-four men.

The pit-head baths at the former Elemore Colliery [57] were built by the Miners' Welfare Fund in 1933 to provide bathing facilities for the men who worked at the Elemore Colliery of Lambton, Hetton and Joicey Collieries Limited. They were one of a number of baths designed by F G Frizzell of the Miners' Welfare Committee [58]. The building's undecorated and almost windowless brick external elevations and its reliance on massing for visual impact are reminiscent of the Dutch Formalist school of architecture, and in particular the work of W M Dudok. The apparent lack of fenestration is explained by the fact that the building is top lit. The structure is compact, making good use of what was a very cramped site. It was designed to accommodate 1,670 men, had sixty-two shower cubicles, locker rooms, and a boot cleaning section. Elemore Colliery closed in the 1970s and the building is at present occupied by a printing firm.

59 *(opposite page) The Avenue Coke and Chemical Plant, Wingerworth, Derbyshire. This was built in the mid 1950s by the National Coal Board as an integrated smokeless fuel and chemical works. Production at Wingerworth ceased in 1992. This photograph is a view along the top of the works' two batteries of Woodhall-Duckham Becker coke ovens. (BB93/1931)*

3

Solid Fuel Manufacture

60 *Whinfield coke ovens, Lamesley, Tyne and Wear. (AA93/998)*

Whinfield coke ovens, Lamesley, Tyne and Wear

NZ 151 581

The manufacture of coke from coal was first carried out in a way similar to the manufacture of charcoal from wood, that is by slow burning in heaps with restricted air contact. Open-heap coking was developed in the 17th century, chiefly to provide fuel for malting. By the end of the 18th century coke was being used for malting, iron working and brass and steel making. In 1709 Abraham Darby successfully smelted iron ore using coke, thereby releasing the iron industry from its reliance upon dwindling timber supplies and paving the way for a massive increase in production and a slashing of prices to its customers. The open-heap method of coking was not, however, suited to some types of Durham coal, which swelled on heating and blocked the spaces in the heap. By the middle of the 18th century it was found that these types of coal could be coked in enclosed ovens. When

Gabriel Jars visited Tyneside in 1765 he saw nine brick-built beehive ovens in three batteries of three (Galloway 1971, 285). The advantages of the beehive oven over open-heap burning were accelerated coking times and improved yields. Open-heap burning could take as long as ten to fourteen days, whereas ovens could carbonise the coal in three to four days. Beehive ovens [**60**] were charged through a loading door in the front or by a hole in the apex of the oven's vaulted roof, the two often being used in combination. After charging, the loading door was sealed, usually with brick. A fire was then lit beneath the oven and the coal roasted until carbonised. Gases produced by the process were vented out through a flue (in primitive ovens the hole in the centre of the domed top of the oven was left open to act as a vent). When the carbonisation was complete, the contents of the oven were quenched with water – usually while still in the oven – and the coke shovelled out through the loading door. Oven vaults were constructed of stone or brick, with refractory brick used increasingly as a lining in the 19th century.

The Whinfield coke ovens were built in 1861 at the Marquess of Bute's Victoria Garesfield Colliery. The bricks from which the ovens were built were made on site and at nearby Lily (later Lilley) Colliery. The coal used in the ovens came from the Victoria seam at the Victoria Garesfield Colliery and the Brockwell seam at the Watergate Colliery (McCall 1971, 56). A narrow gauge railway ran along the top of the batteries of ovens, the trucks (known as 'small tubs') charging the ovens from above (it took four wagons to charge each oven). From the early 20th century the waste heat from the ovens was used to raise steam for generating electricity, which in turn was used by a cuprous oxide plant. This plant, established in 1915, sold much of its output to manufacturers of antifouling paints for ships (McCall 1971, 54). At its peak the works comprised 193 ovens and had an output of 68,000 tons. When it closed in 1958 it was the last plant in the country to be using the beehive oven, its survival to this late date being due to the fact that although the ovens were relatively inefficient by the standards of the time, they did produce coke that was superior in some respects to that produced by more modern plants. Following the closure of the plant a battery of five ovens was preserved by the National Coal Board and in 1973 this became a Scheduled Ancient Monument.

Royston Works of the Monckton Coke and Chemical Company, Havercroft with Cold Hiendley, West Yorkshire

SE 375 122

By-product recovery ovens were developed on the continent in the mid 19th century and were designed to recover gas, tar and other by-products of the coking process. In 1862 the French company Carves and Company acquired the rights to build a type of by-product oven developed by another Frenchman, M Knab, in the 1850s. Twenty years later, in 1882, Henry Simon of Manchester built a battery of Carves ovens at Crook in County Durham. The present coke and chemical works has its origins in the Monckton Main Coal Company, registered in September 1874. In 1875 a brickworks was established on the site and work began on sinking two shafts. In 1879 twelve beehive coke ovens were erected at the colliery, these coming into operation in October of that year. In 1884 several Jameson by-product ovens were built. In 1901 the Monckton Main Coal Company was restructured and renamed New Monckton Collieries Limited. By then the company's Royston coke works comprised 180 beehive ovens and a battery of Simon carves ovens with by-product plant. In the 1920s the coke ovens and by-product plant were transferred to the Monckton Coke and Chemical Company. In the same decade more Simon Carves ovens were erected, to be followed in the 1930s and 1940s by batteries of Becker regenerative ovens. The plant's present ovens date from a rebuild of 1976–9.

Coal is loaded into the ovens from above and the coke ejected through doors in the front (using a ram pushing from behind). The red hot coke is then transported into the tower where it is drenched with water. The forty-two operational ovens each have a capacity of

62 *Pusher side of coke ovens, Royston Works of the Monckton Coke and Chemical Company, Havercroft with Cold Hiendley, West Yorkshire. (AA93/3707)*

15·8 tons of wet coal, which they carbonise into coke in twenty hours at a temperature of 1,300°C. By-product recovery is limited to crude tar and crude benzole (refined off-site). Ammonia is stripped from the gas and concentrated ammonia liquor produced. Gas produced by the ovens is used to maintain the carbonising process and to fire the steam-raising plant. Surplus gas is desulphurised and pressured before being piped to three local industrial customers (including a glassworks).

The past fifty years have seen a massive decline in the amount of coke produced in ovens in Great Britain, this being due largely to the decline of the steel industry. In 1947, 22 per cent of all coal used went to coke ovens, compared with only 9 per cent in 1991–2.

Coalite Works, Bolsover, Derbyshire

SK 462 715

63 *Coalite plant, Bolsover, Derbyshire. (AA 92/7089)*

The best known of all domestic smokeless fuels, Coalite has its origins in a patent for low temperature carbonisation granted to Thomas Parker in 1906. Parker had been approached by environmental groups who were concerned by the smogs that were enveloping towns and cities in the winter months. He responded by developing a low temperature carbonisation process that removed tar and smoke-producing elements from coal, leaving a smokeless fuel that was far easier to light than coke. This fuel was later given the name Coalite.

After a number of false starts, resulting from technical and financial difficulties, the first commercial coalite plant came into production at Barugh, near Barnsley, in 1928. In the following year work began on equipping the plant to produce petrol from by-product coal oil. (Petrol was successfully produced by 'cracking' coal oil at the Killington refinery of Petroleum Refineries Limited.) Four years later the Admiralty completed successful trials of Coalite fuel oil, while the RAF had a squadron of aircraft flying on coal petrol (within twelve months twenty squadrons were using Coalite petrol). In 1932 the company's plants produced 225,680 tons of Coalite smokeless fuel, 26,000 tons of crude oil, 800,000 gallons of petrol and 1,290,000,000 cubic feet of gas. By 1935 the chemical company ICI was also producing petrol from coal oil, and in that year the first petrol pump to serve motorists with petroleum made from coal was officially opened at a garage in London.

The Coalite company's Bolsover works came into production in November 1936, although it was not officially opened until April 1937 [**63**]. When completed it was described as the largest plant of its type in the world. Two years later, in May 1939, a chemical works was opened at the plant. In the years immediately after World War II the market for smokeless fuels was given a boost by legislation to prevent the burning of untreated coal in urban areas. In 1951 ten towns and cities declared smokeless zones. The London smog of December of the following year is believed to have claimed around 4,000 lives, its effects including a nine-fold increase in bronchitis and a four-fold increase in pneumonia. Four years later, in 1956, Parliament passed the Clean Air Act. The year before the Act was passed Coalite built three new batteries of retorts at Bolsover and by 1970 there were twenty-four operational batteries at the works. In 1963 excise duty on home-produced petrol made Coalite oil and petrol uneconomical, and as a result production ceased. In 1986 the company's Askern works (commenced 1928) was closed and production concentrated at Bolsover and Grimethorpe (commenced 1965). Three years later Coalite Smokeless Fuels was formed to operate these two remaining works as a separate trading division of the new parent company, Anglo-United plc (*Coalite 75 Anniversary, 1917–1992*).

4

The Mining Community

64 *Mural, South Hetton, County Durham. The village of South Hetton owes its existence to South Hetton Colliery, the sinking of which began in 1831. In less than a decade a settlement with a population of around 6,000 had sprung up, this being served by eleven places of worship and seven Sunday Schools (Granville 1971, 245). The pit closed in 1983 (its workings merging with those of nearby Murton Colliery), bringing to an end the community's 150 year association with the industry which had called it into being. (AA93/1457)*

65 *Abandoned cottage, Sowood Lane, Whitley Lower, West Yorkshire. (AA 93/1383)*

Abandoned cottage, Sowood Lane, Whitley Lower, West Yorkshire

SE 228 164

Mining villages have tended to be communities standing apart from the rest of the industrial population of the country. The location of mines, and therefore of the mining communities, has been determined by geological conditions. John Holland, writing in *Fossil Fuel* (1835), commented:

> The pitmen in the north of England, reside much less commonly in the towns or villages than in clusters of small houses adjacent to the respective collieries, and forming together little colonies, often more remarkable for the amount of the population, than the neatness or cleanness of their domestic arrangements. (Burton 1975, 196)

Until the middle of the 19th century the majority of pits were not particularly capital intensive ventures and had correspondingly short life expectancies. The scale of the industry was reflected in the housing of the miners.

The single-storeyed cottage on Sowood Lane [**65**] is typical of those built for and by colliers in the Pennine foothills of West Yorkshire in the 18th and early 19th centuries. It comprises a kitchen-living-room and a single bedroom, both of which had fireplaces. The building has walls of coursed sandstone and a roof covered with stone slates. Behind the cottage is a small spoil heap which is marked on the Ordnance Survey map of the area made in 1850 as 'old coal pit', indicating that it was already worked-out by that date. The Parliamentary enquiry into working conditions in the industry published in 1842 described a number of similar cottages in the nearby village of Flockton. In one such two-room cottage Joseph and Anne Charlesworth lived with their four children. Joseph and his eldest son George, aged nine, both worked at a local colliery. The Charlesworth's home was furnished with two beds, a clock, chairs, two or three tables, drawers, a delph case, cradle, and the cages of Joseph's six pet larks. The inspector described Joseph as 'a good and regular worker, but extremely negligent of his religious duties, spending the Sabbath in smoking and feeding his birds' (Parliamentary Papers 1842, 217–19).

66 *Reform Row, Wath Road,*
Elsecar, South Yorkshire.
(AA93/1014)

Elsecar village, Hoyland Nether, South Yorkshire

SE 39 00

By the early 19th century a number of philanthropic coal owners were providing higher than average quality housing for the miners working in their pits. One such was Earl Fitzwilliam, whose village of Elsecar was regarded by contemporaries as a model industrial community [**66** and **67**]. The Fitzwilliams were paternalistic employers who provided the workers at their ironworks and pits with a wide range of benefits, including widows' pensions, allowances to injured miners, and medical attention in the event of accident. A report of 1845 noted that in contrast to the conditions found in many mining settlements, the housing at Elsecar was:

... of a class superior in size and arrangement, and in conveniences attached, to those belonging to the working classes. Those at Elsecar consist of four rooms and a pantry, a small back court, ash-pit, a pig-sty, and a garden; the small space before the front door is walled round, and kept neat with flowers or paving stones; a low gate preventing the children from straying into the road. Proper conveniences are attached to every six or seven houses, and kept perfectly clean. The gardens, of 500 yards are cultivated with much care. (Mee 1976, 49)

In 1853 a lodge was opened which provided accommodation for twenty-two single men. The symmetrical façade, pediment and projecting central bays give the building the outward appearance of a Georgian country house [**68**]. Each lodger had his own room, an unusually enlightened arrangement for the period and a sharp contrast to the austere barracks provided by some employers. The lodge also contained the first bath and hot water geyser in the village. The building later became the 'bun and milk' club, a kind of temperance working men's club. After lying empty for a number of years it was converted into flats in the late 1980s.

67 *Datestone, Reform Row, Elsecar, South Yorkshire. (AA93/1446)*

68 *Former miners' hostel, Fitzwilliam Street, Elsecar, South Yorkshire. (AA93/1059)*

69 New Bolsover Colliery Village, Bolsover, Derbyshire. (AA92/7092)

New Bolsover Colliery Village, Bolsover, Derbyshire

SK 465 704

The dramatic expansion of the coal industry in the second half of the 19th century meant that a growing number of workers needed to be attracted to the coalfields. Between 1850 and 1914 the number of persons employed in the industry grew from 200,000 to around 1,200,000. In order to obtain men for new collieries, many coal companies were obliged to provide housing close to their pits. The provision of housing also allowed the companies to exercise a degree of social control over their work-forces. This was very much the case at the Bolsover Colliery Company's village of New Bolsover, work on which began in 1891 [**69**].

The village, designed by the partnership of Brewill and Bailey of Nottingham, was built as a model community. The houses were arranged in a double horseshoe around a large green. The company provided both two and three-storeyed houses for its workers, the latter having additional bedrooms in attics lit by dormer windows. A tram track ran between the backs of the houses, the purpose of which was to transport the men to the pit, the free coal from the pit, and to carry away the contents of the ash heaps and privies. In addition to the terraced rows, there were also semi-detached villas for colliery officials and white-collar staff. The village had a school, institute, co-operative store, Methodist and Anglican churches and an orphanage. There were no public houses and the bar staff at the institute were prohibited from serving more than three glasses of beer to each worker in an evening. On the edge of the village the company provided a cricket ground, allotments and pigsties. The inhabitants were watched over by a policeman employed by the company, misconduct being punished by fines or dismissal (Haigh 1989, 10).

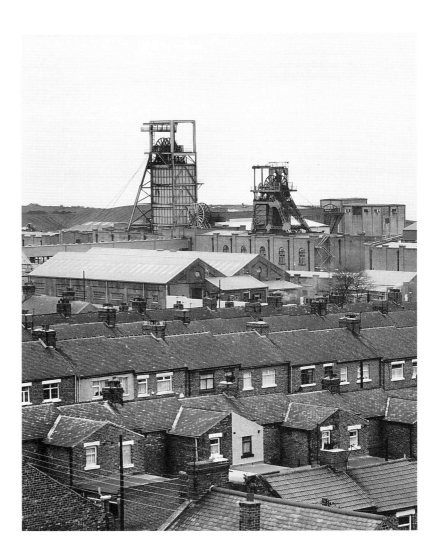

70 *Terrace housing and pit, Easington Colliery, County Durham. (AA92/7043)*

Easington Colliery, County Durham

NZ 43 44

The increasing size of individual pits led to the building of mining communities on an ever larger scale, the small clusters of miners' cottages which had characterised miners' housing earlier in the 19th century giving way to long rows of terraced houses. The housing provided by many coal owners was often built as cheaply as possible, giving miners' housing a poor reputation. Moreover, pit villages, in common with rural settlements, tended to suffer from an absence of drains and from problems of obtaining adequate supplies of fresh water. In 1892 the sanitary committee of Durham County Council went so far as to declare that the coalfield's housing problems required urgent action (Colls 1987, 264). In the same year the *North Eastern Daily Gazette* ran a series of articles called 'Homes of the

Pitmen', which drew attention to the poor housing conditions in the county's mining communities (Colls 1987, 265).

The village of Easington Colliery was built in the first decade of the 20th century, to serve a pit then being sunk by the Easington Colliery Company [**70**]. In many respects it is typical of the company housing of the period, and was better than much of the older miners' housing in the county. The houses were laid out as uniform rows on a grid plan and no open space was left between the village and the pit it served. The brick-built, two-storeyed rows with their Welsh slate roofs are indistinguishable from much other industrial housing of the period. The degree of regimentation imposed by the village's grid plan was reflected in the fact that the streets were given numbers rather than names, starting with First Street. However, the company did later relent and gave each row a name, the rows of each block starting with the same letter of the alphabet (eg, Abbot Street, Ayre Street, Allen Street, etc). By the time the housing at Easington Colliery was completed, attitudes to the design of workers' housing were already beginning to change, with more thought being given to the planning of new communities and more consideration to the quality of the environment in which the families of miners were to live.

71 *Aerial view of Woodlands Colliery Village, Adwick le Street, South Yorkshire. (SE5307/3)*

Woodlands Colliery Village, Adwick le Street, South Yorkshire

SE 53 07

The early years of the 20th century saw the first attempts to apply the principles of the 'garden village' to the design of coalmining communities. One of the first was Woodlands village, built from 1907 to 1909 for the Brodsworth Main Colliery Company Limited. Sir Arthur Markham, chairman of the company, wished to see the work-force of the new pit well housed in a model village [**71** and **72**]. His sister, Violet Markham, later said of Sir Arthur that he believed it intolerable 'that miners, or any other group of men, should live and work in bad conditions; that their wages should fall below a proper standard of life and the houses in which they lived should be bereft of all decent amenities – houses unfit to bear the name of home' (Anon 1955, 44).

The architect chosen for the Woodlands project was Percy Houfton of Chesterfield, a trained mining engineer who had turned his attention to the designing of model workers' housing. Prior to Woodlands he had produced a number of schemes for colliery housing, including Cresswell Model Village, Derbyshire, and in 1905 had been awarded the £100 prize for the best cottage at the exhibition of cheap cottages at the Garden City, Letchworth. The houses at Woodlands are brick-built in blocks of two, three, four and five. Some have the brick exposed completely, some have roughcast first floors and some are rendered all over. Houfton's houses are in the Arts and Crafts tradition, some having large gables reminiscent of the work of C F A Voysey [73]. A variety of types of accommodation was provided, from houses with large living-room, scullery and three bedrooms to ones with parlour, kitchen, scullery, three bedrooms and a fitted bathroom. Among the houses provided with fitted baths, some had the bath in a first-floor bathroom while others had it in a ground-floor scullery immediately next to the back door. The scheme also made provision for Anglican and Wesleyan churches, a Primitive Methodist chapel, schools and a co-operative store, these being allocated plots in the centre of the village.

The sinkers working at Brodsworth Main reached the Barnsley Bed sooner than was expected (October 1907), with the result that the company pressed for the village to be finished as quickly as possible. The result was that the number of house types was reduced in order to speed up the building campaign, 500 houses being built and in occupation in just over a year. In 1910 the *Architectural Review* conceded that Woodlands showed 'some attempt to realise Garden Village and Town Planning principles' (Anon 1910, 184), but lamented that the speed with which the latter part of the village was built had 'involved some loss in the interest and quality of the later work' (Anon 1910, 186). The *Victoria County History*, published two

72 *36–9 Woodlands Green, Woodlands Colliery Village, Adwick le Street, South Yorkshire. (AA93/1715)*

73 *Housing built for mine officials, Whitwood, Castleford, West Yorkshire. The houses were designed by C F A Voysey for the Coal Company of Henry Briggs and Son. (BB93/21393)*

years later, was more generous in its praise:

> A pleasing feature of the modern colliery proprietor is the desire to provide better accommodation in regard to housing than has hitherto been attempted. At Dinnington, Brodsworth, Bentley and Maltby model villages are being erected, instead of the long depressing rows of houses so commonly met with in the Barnsley and Wakefield districts. (VCH 1912, 326)

Aylesham Village, Nonnington, Kent

TR 23 52

When the Royal Commission on the Coal Industry reported in 1926 it was highly critical of the general standard of the housing provided by the majority of coal companies:

> The housing conditions of colliery workers, like almost all else connected with the industry, show great diversities. They are often very bad – many of the old villages consist of poorly constructed cottages, small and frequently overcrowded, with sanitary arrangements primitive and inadequate, the aspect of the villages being drab and dreary to the last degree. (*Report of the Royal Commission on the Coal Industry* 1926, 199)

The President of the Mining Association of Great Britain, Evan Williams, accepted that there were houses 'built long ago in connection with pits that are nearing the end of their life and which do not conform to the ideals of 1927', but went on to argue that no one could accuse the colliery owners 'of building otherwise than in accordance with the accepted standard of the day', a fact which was illustrated 'by the garden villages which have been built near pits

74 *Block of four miners' houses, Ratling Road, Aylesham Village, Nonnington, Kent. (BB93/8678)*

recently sunk in all parts of the country' (*The Iron and Coal Trades Review* 1927, 6).

The 1920s did indeed see the building of a number of model mining communities, mainly in the expanding coalfields of South Yorkshire, North Derbyshire, Nottinghamshire and Kent. By the middle of the decade the coal companies operating in these areas were allocating around 40 per cent of their investment in new pits to housing schemes. In addition to the houses built by individual colliery undertakings, twenty-four colliery companies from South Yorkshire and the East Midlands collaborated in establishing the Industrial Housing Association, which, by the end of 1925, had built 8,075 houses. The Association's houses were of high quality, being built in blocks of two or four and having three or four bedrooms and a bathroom.

In 1925 the British Empire Exhibition at Wembley had included a colliery housing exhibit, 'designed to show the very marked improvement which has been made during the last 30 years or so' (*The Colliery Guardian* 8 May 1925). One of the features of the model house to which attention was drawn was the provision of a bathroom on the ground floor:

This is approached off the scullery, which is the best arrangement for a collier or any other man coming home from work dirty. He is so enabled to enter the house by the back door and go straight to the bath without passing through any of the better part of the house. There is in fact no need for 'pit muck' ever to go beyond the scullery door. (*The Colliery Guardian* 8 May 1925)

The villages built for the new Kent coalfield were noted for the quality of their housing. Among these was the village of Aylesham, built from 1926 to 1927 by Pearson and Dorman Long and Company to house miners employed at the company's Snowdown Colliery (opened 1912). The 'new town', as it was called in the early years, was designed by J Archibald, C T F Martindale and Sir Patrick Abercrombie. It has a Beaux-Arts plan, with roads curving symmetrically around a central grassy square. A variety of types of housing was provided, their size and facilities reflecting the position of the occupant within the company hierarchy. Most houses were built in blocks of two or four [**74**], the smaller dwellings having a living-room, three bedrooms and a scullery which doubled as a bathroom, while the larger comprised a parlour, kitchen, three bedrooms and a fitted bathroom. In addition, the village was provided with shops, school, cinema, medical centre, church, chapel and a branch of the International Miners' Mission (see pp 98–9).

Dorman Long was badly hit by the depression in the coal and steel industries of the 1920s and early 1930s, being unable to declare a dividend on ordinary shares between 1922 and 1935. Throughout the 1920s the company had benefited from a Treasury guarantee of much of the money it invested in the Kent coalfield. However, in 1929 it decided to scale down investment and not to complete Aylesham as originally planned, a scheme that would have resulted in the creation of a small town of 15,000 inhabitants. Difficult mining conditions had hampered the development of the coalfield, and in 1930 Kent's pits had produced only 1,300,000 tons. Snowdown Colliery, Aylesham's *raison d'être*, closed in 1987 and Betteshanger Colliery in 1989, the closure of the latter bringing to an end the story of the county's short-lived coalmining industry.

75 *1–3 Laburnum Close, Townville, Pontefract, West Yorkshire. (AA93/1635)*

NCB housing, Townville, West Yorkshire

SE 4552 2447

At Nationalisation the NCB inherited some 140,000 colliery houses from the private companies, 37 per cent of which were classified as being in 'poor' condition and only 34 per cent as being 'reasonably modern and in fair condition' (Ashworth 1986, 541). Some new housing was built in mining communities by the Ministry of Health in the early 1950s, but this did not meet the shortfall and in 1952 the NCB set up the Coal Industry Housing Association (CIHA). The Association aimed to build 19,842 houses, mostly in 'manpower deficiency' regions of Yorkshire and the East Midlands. Most of the planned houses were completed by 1955 and ten years later a total of 24,070 houses had been built.

The NCB housing scheme at Townville [75] was the work of the architectural practice of Antony Steel and Owen, although the houses themselves were designed by the firm of Messrs Wates Limited. The plans for the estate were submitted by the regional surveyor of the Board on behalf of its Housing Association in September 1952. The original plan was for 300 houses, but this was later increased to 312. The site was laid out by Hadsphaltic Construction Limited and the houses were built by Shepherd and Son. The houses were of three types: the EM, EM2 and EM3. All had bathrooms, these differing from their pre-war counterparts in being located on the upper rather than the ground floor. The switch to upper floor bathrooms was the result of the introduction of pit-head baths, the effect of which was to obviate the need for miners coming off shift to bathe at home. The houses had reinforced-concrete frames and concrete outer walls, the inner walls being of brick and breeze blocks. The picket fences around each property were also of reinforced concrete.

Rationalisation of the industry in the 1960s led to a rapid shrinking of the work-force, and by the end of the decade the NCB was reducing its housing stock by a combination of demolition and sale (either to local authorities or sitting tenants). By 1972 the number of NCB houses had fallen to 103,498 (Ashworth 1986, 544), 32 per cent of which were occupied by retired workmen and widows. In 1976 the decision was taken to sell off all NCB workers' housing as quickly as possible, the only exceptions being houses required for operational reasons (such as rescue brigade houses). In 1977 all CIHA housing was transferred to the NCB. By March 1983 the number of NCB houses had fallen to 36,000. The housing at Townville was transferred to Wakefield Metropolitan District Council in the 1980s and demolished from 1992 to 1993.

Joicey Aged Miners' Homes, Shiney Row, Houghton le Spring, Tyne and Wear

NZ 323 529

76 *Joicey Aged Miners' Homes, Houghton le Spring, Tyne and Wear, built by the Durham Aged Miners' Homes Association in 1906. (AA93/1082)*

Many coalminers lived in houses owned by the coal companies for which they worked. The result was that men could be evicted when age or injury made them unable to continue working. They could also be evicted for striking, although this became less common after the end of the 19th century. If a retired miner had two or more unmarried sons working at the colliery and living at home, then he might be allowed to keep his company house. It was expected that retired miners would live with married sons or daughters, the only alternative being the workhouse. It was not uncommon for widows of miners killed at work to be allowed to continue to occupy their company house for up to two years before being evicted, an exception being made if they had sons working at the colliery and living at home. In short, the disabled, old men and widows were all liable to be evicted from their homes in order to make way for younger men with families.

Around the end of the 19th century a number of organisations were created to build and administer homes for retired miners. The largest of these was the Durham Aged Miners' Homes Association, which was established in 1897 by the Durham Miners' Association. The funds were obtained by a non-compulsory weekly levy on miners working in the Durham coalfield. Within ten years the Association had provided no fewer than 277 dwellings. Most of the houses built were single storeyed in terraced rows. A typical dwelling comprised a kitchen-living-room, scullery and bedroom. The shrinkage of the industry from the 1930s and the resulting reduction of income from the levy led to growing financial difficulties. In 1980 the Association became a registered housing association, thereby becoming eligible for Housing Corporation funding.

The Joicey Aged Miners' Homes at Shiney Row [76], Houghton le Spring, built in 1906, are typical of the housing provided by the Durham Aged Miners' Homes Association. They are named after one of County Durham's principal mine-owning families. The Joicey family's involvement with the industry began in the early 19th century, when James Joicey trained as a coal viewer and mining engineer at South Hetton. In 1837 he went into partnership with his brothers John and Edward. In 1881 the firm came under the management of John's nephew, also called James, in whose time it took over both the Hetton and Lambton Collieries. When James, later Lord Joicey, died in 1936 he was one of the wealthiest mine owners in the country, leaving a personal fortune of £1,500,000.

Houses of Rest for Miners, Park Drive, Hucknall, Nottinghamshire

SK 534 484

77 *Houses of Rest for Miners, Park Drive, Hucknall, Nottinghamshire, designed by Sir Reginald Blomfield for Sir Julien Cahn. (AA93/1351)*

These six homes for retired miners were designed by Sir Reginald Blomfield and built in 1925 by Sir Julien Cahn. They are an example of private philanthropy rather than collective self-help. The homes, which are in Blomfield's 'Wrenaissance' style, comprise three single-storeyed blocks in a half-butterfly plan. They are brick-built with ashlar dressings, and have hipped roofs. The central two bays of each block break forward and have a pediment over. The central block, which is surmounted by a cupola, has a centrally positioned plaque commemorating Cahn's benefaction. Cahn's obituary in the *Nottingham Guardian* (28 September 1944) commented that 'No man was more liberal with his wealth for deserving causes', these including the provision of a new orthopaedic clinic in Hucknall [**77**], a fully equipped maternity hospital at Stourport and the Pay-Bed Wing at Nottingham General Hospital.

78 *Primitive Methodist Chapel, Railway Street, Hetton le Hole, Tyne and Wear. (AA93/1434)*

Primitive Methodist Chapel, Railway Street, Hetton le Hole, Tyne and Wear

NZ 355 479

The Primitive Methodists broke away from the Wesleyans around the year 1810, one of the principal causes for the schism being an argument over what social historians have called 'religious democracy'. Wesleyanism was centralised, hierarchical and made a clear distinction between preacher and layman, whereas Primitive Methodism laid strong emphasis on lay preaching, including the right of women to preach. The Primitive Methodists also took up the 'camp meeting', a device invented by American frontier evangelists in the late 18th century. These meetings were held in the open air and often included preaching of such fervour that it could induce collective hysteria and mass conversion. The founder of Primitive Methodism, Hugh Bourne (a Staffordshire millwright), was later to recall that

'Our chapels were the coal-pit banks, or any other place; and in our conversation way we preached the Gospel to all, good or bad, rough or smooth' (Thompson 1968, 436). From the beginning the new sect's membership was predominantly working-class. Primitive Methodism reached the Durham coalfield in the early 1820s and in 1823 the Primitive preacher Thomas Nelson was able to report:

> A very blessed and glorious work has gone on for some time in Sunderland and the neighbouring collieries. In Sunderland and Monkwearmouth ... we have nearly 400 members. In Lord Steward's [Stewart's] and Squire Lambton's collieries we have near 400 more ... Indeed, the Lord and the poor colliers are doing wondrously. (Colls 1987, 150)

The chapel at Hetton le Hole was built in 1858, probably by miners of the Hetton Colliery [78]. The front wall is of coursed sandstone and the sides and rear of limestone rubble. The walls are said to incorporate stone sleeper blocks from a disued pit wagonway. The building has an imposing street elevation in the centre of which is a massive tripartite window lighting both ground and first floors. This window has semicircular arched heads and a heavy transom which bears the inscription 'PRIMITIVE METHODIST CHAPEL'. There are two entrances, each with a semicircular headed window above it. Both these entrances are approached by stone steps with iron railings. The elevation is surmounted by a gable of rather unusual appearance, having long sloping parapets with raised copings, the ends of which are defined by large end blocks with arched tops.

Church of St Peter, New Fryston, West Yorkshire

SE 453 268

In 1844 John Spencer of the Phoenix Iron Works, West Bromwich, developed machinery to manufacture corrugated-iron sheets, the purpose of the process being to give thin sheets of wrought iron greater strength. By the 1860s galvanised corrugated-iron sheets were being used for roofing and cladding a variety of types of building (Emery 1990, 59). Among these were 'tin' churches and chapels – wooden buildings with a roof covering and wall cladding of corrugated iron. The advantages of this method of construction were that it was quick, cheap (costing approximately £1 per seat provided) and allowed the building to be dismantled and re-erected on another site if and when necessary. Buildings of this type became a common sight in the coalfield communities of England, the majority being built by the Church of England, Catholics, Methodists and Baptists.

The former Church of St Peter was erected by the Church of England around 1895 to serve the new mining community of New Fryston. The building was rectangular in plan, with transepts that projected slightly and a small west porch. An early photograph of the building shows that at least some of the windows originally had Y tracery [**79**]. The details of construction, and the tension braced

collar trusses in particular [**80**], suggest that it was built by W Harbrow, a London firm specialising in corrugated-iron buildings. In 1902 Harbrow advertised 'New and second-hand Churches, Chapels, Mission and School Rooms, Cottages, and Keeper's Huts, Stables, Coach-houses, Farm Buildings, Sheds for Manufacturing Purposes, &c' (Emery 1990, 60).

Fryston Colliery closed in 1985, since when the community has contracted to a population of around one hundred, housed in a few rows of terrace houses. The Church of St Peter was demolished in 1992.

80 *Interior of former Church of St Peter, New Fryston, West Yorkshire. (BB93/5923)*

81 *International Miners' Mission, Burgess Road, Aylesham, Kent. (BB93/8655)*

International Miners' Mission, Burgess Road, Aylesham, Kent

TR 2402 5270

The International Miners' Mission was founded in 1906 by Richard Glynn Vivian, a wealthy industrialist and brother of Lord Swansea. He was both inspired and assisted by Herbert Voke, initially his valet but later his private secretary and spiritual counsellor, and Colonel James Philips, the superintendent of the Union Street Mission in Brighton. The influence of the latter man in particular is clear from Glynn Vivian's own account of the establishment of the International Miners' Mission:

Just as I began to go blind, I saw at Brighton the excellent work of a Mission under Col. Philips, and the idea seized me to form a similar one at

Hafod, just outside Swansea, where my company owned a large mine, so that our workmen and colliers could be reached for Christ. Under God's guidance this grew in my mind to the idea of forming missions for miners, colliers and workmen all over the world. (Anon nd)

The organisation founded by Glynn Vivian went on to establish five missions in Great Britain and others in Austria, Chile, France, Germany, Italy, Japan, Malaya, Nigeria, Portugal, Spain, Zaire, Zambia and Zimbabwe. The mission at Aylesham was opened in 1928 and soon proved to be a popular religious and social focal point for the new mining community [81]. The mission hall was built with money originally intended to be used to finance a mission in Russia, a venture which was abandoned when the new Communist regime banned Christianity. It closed following the shutting of Snowdown Colliery in 1987, a rule of the organisation being that it could not serve communities that did not have working mines.

*82 Church of St Paulinus, New
Ollerton, Nottinghamshire.
(AA93/929)*

Church of St Paulinus, New Ollerton, Nottinghamshire

SK 675 683

The village of New Ollerton was built by the Butterley Company, the first tenders for housing being invited in July 1922. It was one of a number of colliery villages built in the 1920s and 1930s to house the labour force that was needed to work the seven large pits sunk in the expanding Nottinghamshire coalfield in the 1920s, others including Blidworth, Clipstone and Edwinstowe. In 1926 Butterley's directors decided that the company should build a church in the village as 'a cathedral for the new coalfield' (Waller 1979, 72). Sir Giles Gilbert Scott produced plans for a church and vicarage at the company's request, but these were rejected and Scott dismissed as architect. The church was designed by the practice of Naylor, Sale and Woore and built at a cost of £8,000 (£5,000 of which was contributed by the company). The Butterley Company also gave contributions towards the building of Baptist, Methodist and Roman Catholic churches, and to a Salvation Army citadel.

The foundation stone of the church was laid by the Bishop of Southwell on 13 March 1931. The building is of brick construction in a plain Romanesque style with gabled, sprocketed plain tile roofs [82]. The gables have diaper work, cogged eaves and tile kneelers and corbels. The nave, chancel and sanctuary are all under a continuous roof, while transepts which double as porches project from the north and south elevations. The nave roof has king-post trusses with double tie-beams and a coved wallplate carved with an interlaced pattern. The church's fittings include a cruciform brick font and a canted brick pulpit with leaded top and bronze handrail.

83 *Miners' Memorial Chapel, All Saints' Parish Church, Denaby, South Yorkshire. (AA93/1323)*

Miners' Memorial Chapel, All Saints' Parish Church, Denaby, South Yorkshire

SK 502 994

The Miners' Memorial Chapel in Denaby, completed in 1987, was built as a memorial to those who worked in the collieries of the area [83]. The walls are built of bricks salvaged from Cadeby, Hickleton, Kilnhurst, Manvers and Rossington Pits, and from the former power station at Mexborough. The pit wheel, from Cadeby Colliery, is 19 ft in diameter and weighs 5 tons. It was this wheel that raised the last coal mined at the colliery in 1986. The archways that frame the altar are supported by steel 'gateways', also from Cadeby. The altar itself is a glass-sided case containing a 1 ton block of coal from Manvers Main Colliery.

84 *National Union of Mineworkers' offices, Victoria Road, Barnsley, South Yorkshire. (AA92/7018)*

National Union of Mineworkers' offices, Victoria Road, Barnsley, South Yorkshire

SE 343 070

The first attempt at a national union for workers in the coal industry was the Miners' Association, founded at Wakefield in November 1842. After some early successes the Association went into decline, going out of existence in 1849. In the decade that followed a number of regional unions were established, one of which was the South Yorkshire Miners' Association (founded 1858). The present NUM offices in Barnsley were built to provide that union with permanent offices and a meeting room [**84**]. When opened in November 1874 it was one of only a handful of purpose-built trade union headquarters. The opening was marked by the Miners' National Association Conference – a pressure group for legislative reform – being held in the building in the same month. Welcoming the delegates, the South Yorkshire miners' secretary John Normansell, declared that it gave him 'great pleasure indeed to say that you are welcome into a house that is built by your fellow miners at their own cost and expense in every way'. (NUM Souvenir Brochure 6)

The building was designed by Wade and Turner of Barnsley, who produced imposing elevations which freely combine gothic with French Empire. A meeting hall was added in 1912, the interior of which is richly decorated. The windows of this hall have stained glass medallions depicting scenes from colliery life, a number of which are reproduced in this book. The monument in front of the building commemorates leading figures in the early history of the South Yorkshire Miners' Union, and in particular John Normansell (general secretary from 1864 to 1875). Normansell was largely responsible for the building of the headquarters and died within a year of their completion.

In 1889 a campaign by coalminers for a ten hour day led to the founding of the Miners' Federation of Great Britain. This organisation was supported by most of the regional miners' unions, with the exception initially of those of Durham, Northumberland and South Wales. On 1 January 1945 the forty-one constituent bodies of the federation were replaced by a single union, the National Union of Mineworkers. In 1988 the national headquarters of the NUM moved from London to Sheffield [**86**].

85 *Stained glass depicting the building as it looked in 1912, NUM offices, Barnsley, South Yorkshire. (AA93/3443)*

86 *National Union of Mineworkers' headquarters, Holy Street, Sheffield, South Yorkshire. (AA93/1343)*

*87 Esh Winning Miners' Memorial
Hall, Brandon with Byshottles,
County Durham. (AA93/1084)*

Esh Winning Miners' Memorial Hall, Brandon Road, Esh Winning, Brandon with Byshottles, County Durham

NZ 193 419

A number of miners' memorial halls were built in coalfield communities after World War I. The hall at Esh Winning [**87**] was designed by J A Robson and built by E R Davison and Sons of Blaydon at a cost of £10,024 (Emery 1992*b*, 87). It was completed in 1923 and is a late example of Edwardian Baroque, the same style as chosen for the Miners' Association headquarters in the nearby City of Durham, built from 1913 to 1915. The walls are of light red engineering brick and yellowy-brown terracotta. The centre and end sections break forward, the former being surmounted by a segmental pediment and the latter by open triangular pediments. This large and impressive building, now empty and derelict, once offered a wide range of facilities, including a cinema, concert hall, swimming bath, billiard room, games room and library. The cost of building and fitting out the hall was met partly by a contribution of 3*d* a week from local miners and partly by a generous contribution from the colliery company of Pease and Partners, 543 of whose employees had died in the war.

88 *Miners' Welfare Institute, Kibblesworth, Lamesley, Tyne and Wear. (AA93/1356)*

Miners' Welfare Institute, Barrack Terrace, Kibblesworth, Lamesley, Tyne and Wear

NZ 567 244

The Miners' Welfare Commission was fortunate in that the legal definition of its purpose ('the social well-being, recreation and conditions of living of workers in and about coal mines') was so broad that it enabled the Fund to provide a wide range of welfare provision 'ranging from convalescent homes to bagpipes for a band of kilted miners; from glass eyes and artificial limbs to institutes and community centres' (*Mining People* 1945). In particular, the Miners' Welfare Institutes and recreation grounds provided a focal point for the communities they served. By 1945 the Fund had given financial assistance to 1,500 schemes at a cost of £5,950,000. The 'Welfare' varied from 'a small hut with a billiard table' to 'a building of conspicuous architectural merit with provision for leisure-time activities of members of the community of all ages and both sexes, such as dances, concerts, plays, discussions, gymnastics and the cinema' (*Mining People* 1945).

Among the institutes built with money provided by the Fund was Kibblesworth Institute, County Durham [**88**], the foundation stone of which was laid by the Duchess of York in 1936. It was built to serve the work-force of Kibblesworth Colliery, owned at the time by the firm of Dorman Long and Co. The institute replaced an earlier structure described at the time as a 'Tudor brick building', the Fund's report of 1937 commenting that 'in its elevational treatment the character of this old work and of the village has been preserved by the architect, Mr Edwin Lawson, ARIBA' (Miners Welfare Fund 1937, 52). The building is indeed an interesting and thoughtfully executed example of Tudor revival, its size, large chimney and mullion and transom windows suggesting a domestic rather than institutional function. Kibblesworth Colliery ceased production in 1974 and was demolished in 1975.

90 *Practice night for the Sharlston Colliery Band, Sharlston, West Yorkshire. (AA93/3714)*

89 *Sharlston Miners' Welfare, Sharlston, West Yorkshire. (AA93/3712)*

Miners' Convalescent Homes, Bispham near Blackpool, Lancashire, and Skegness, Lincolnshire

Bispham SK 394 308   Skegness TF 658 574

In 1924 the Miners' Welfare Fund stated its preference for converting existing country houses for use as convalescent homes, arguing that this was cheaper than providing purpose-built accommodation. By 1927 convalescent homes had been or were in the process of being provided for the miners of Cannock Chase, Derbyshire, Lancashire, North Staffordshire, South Yorkshire and the Mansfield area of Nottinghamshire. Of these, only the homes at Blackpool [**91**] and Skegness [**92**] were purpose built, the others being large houses converted for the purpose. The Lancashire, Cheshire and Derbyshire miners went against the advice of the Fund and opted for purpose-built homes on the grounds that no suitable accommodation was available in their chosen locations, these being the popular seaside resorts of Blackpool and Skegness, respectively.

Work on the two homes went on through the protracted coal strike of 1926, and both were completed in 1927. The Blackpool home was opened in June 1927 by the Prince of Wales. It had grounds of 7½ acres, cost £160,000 to build and equip, and provided beds for 132 men. The Skegness home occupied a 4 acre site (donated by the Derbyshire Miners' Association), cost £60,000 to build and equip, and had beds for 124 male and thirty-four female patients. The two buildings are very different in appearance, Skegness (by F H Broomhead of the firm of Percy B Houfton and Co) being designed in restrained neo-Georgian style and Blackpool (by Bradshaw Gass and Hope) in a style which owes much to French Renaissance.

92 *Derbyshire Miners' Convalescent Home, Winthorpe Avenue, Skegness, Lincolnshire. (AA93/1438)*

93 *Castlepark Bandstand, Flatt Walks, Whitehaven, Cumbria. (AA93/1318)*

Castlepark Bandstand, Flatt Walks, Whitehaven, Cumbria

NX 968 178

The Welfare Fund provided a large number of recreation grounds for mining communities, as well as giving grants to local authorities to establish such facilities or improve existing ones. The bandstand at Castlepark, Whitehaven [**93**], was designed by J A Dempster of the Miners' Welfare Committee and built in 1934. It cost £855 to build, £755 of which was contributed by the Fund. The structure was almost entirely of reinforced concrete, the only exceptions being the sliding glazed partitions at the sides. Accommodation for artistes and storage was provided in a basement below the platform.

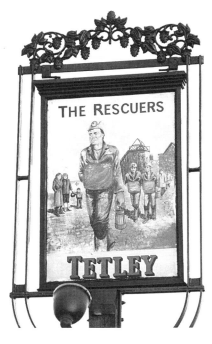

94, **95**, **96**, **97** *(left to right) Four Yorkshire pub signs which celebrate the mining industry: The Askern, Instonville, Askern, South Yorkshire; the Miners' Arms, Aberford Road, Garforth, West Yorkshire; the Goldthorpe, Doncaster Road, Goldthorpe, Dearne, South Yorkshire; and The Rescuers, Lofthouse Gate, Normanton, West Yorkshire (a name given following the Lofthouse mining disaster of 1973). Of the four, only the Goldthorpe served a community with a working pit at the time of writing. (AA93/1344, AA93/1004, AA93/1006, AA92/7076)*

5

Mines Safety

Oh, let's not think of tomorrow, lest we disappointed be,
Our joys may turn to sorrow as well all may daily see,
Today we may be strong and healthy, but soon there comes a change,
As we may see from the explosion that has been at Trimdon Grange.

The Ballad of Trimdon Grange Tommy Armstrong (1881)

98 *Detail of the Trimdon Grange Memorial, Trimdon, County Durham. On 16 February 1882 seventy-four men and boys were killed as a result of an explosion at Trimdon Grange Colliery. Among the dead were members of a rescue party who entered the workings from those of nearby Kelloe Colliery. (AA93/960)*

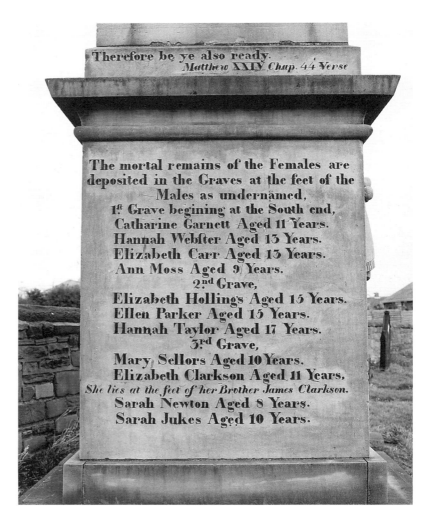

99 *Huskar Pit Disaster Memorial, Silkstone, South Yorkshire. (AA93/1061)*

Huskar Pit Disaster Memorial, Silkstone, South Yorkshire

SE 291 059

On 4 July 1838 twenty-six children were drowned when torrential rain flooded Huskar Pit. The dead included seven boys under the age of ten and a girl of eight [**99**]. On 1 July 1842, almost four years to the day after this disaster, the Report of the Commission on Conditions in Mines was published. The report caused an outcry by revealing to an appalled public the conditions in which young children worked in the mines. The outcome was Lord Ashley's Mines Regulation Act of 1842, which prohibited boys under ten and females from working underground. Twelve years later the Coal Mines Regulation Act of 1860 raised the minimum age from ten to twelve.

Hartley Colliery Disaster Memorial, St Alban's Churchyard, Earsdon, Tyne and Wear

NZ 320 726

On 16 January 1862 the beam of the New Hartley Colliery's pumping engine broke in two and half of it crashed down the shaft trapping 204 men and boys in the workings below [**100** and **101**]. Three days later an observer described the scene at the colliery: 'All around the pit-heap the watch-fires are still burning, and the patient waiters, looking more solemn and despairing than ever are clustered near them. They have withstood the severity of one of the bitterest nights we have had during the present winter, and as yet they show no signs of giving in' (McCutcheon 1963, 54). Three more days passed before the rescuers, a party of sinkers, got beyond the blockage in the shaft only to find that there were no survivors, the trapped miners having suffocated. The disaster had a devastating effect on the community, it

101 *Detail of Hartley Memorial. (AA93/1330)*

being said 'that there was scarcely a house in the village in which the calamity was not felt and from which one or more coffins were not brought forth' (Fynes 1873, 178). Indeed, the dead were too numerous for the old churchyard at Earsdon and a neighbouring field had to be taken over for the purpose of burial.

The Hartley disaster had the effect of focusing public attention on the issue of mines safety and 'excited more inquiry into the nature of the work of miners than any previous casualty has done' (McCutcheon 1963, 115). The principal cause of the great loss of life was identified as the pit having only one shaft, thereby leaving the miners no means of escape. Parliament responded swiftly with an Act to Amend the Law Relating to Coal Mines being passed in August 1862. The Act stated that it was unlawful for the owner of a new mine to employ men to work a mine that did not have at least two shafts or outlets separated by natural strata not less than 10 ft in breadth. Owners of existing mines were given until 1 January 1865 to provide a second shaft or outlet.

102 *Barnsley Oaks Disaster Memorial, Ardsley, South Yorkshire. (AA93/274)*

Barnsley Oaks Disaster Memorial, Ardsley, South Yorkshire

SE 352 058

As the 19th century progressed shafts were driven deeper, making it increasingly difficult to ventilate workings adequately. The failure of ventilation and safety precautions to keep pace with the growing scale of the industry led to a series of major colliery disasters. In South Yorkshire, attempts to work the gaseous seams of the Barnsley area resulted in a number of catastrophic explosions. Between 1840 and 1860 more than 600 local men and boys were killed in this way. The most calamitous gas explosion in the history of the English coal-mining industry occurred at Firth Barber and Company's Old Oaks Colliery near Barnsley on 12 December 1866 [**102** and **103**]. The

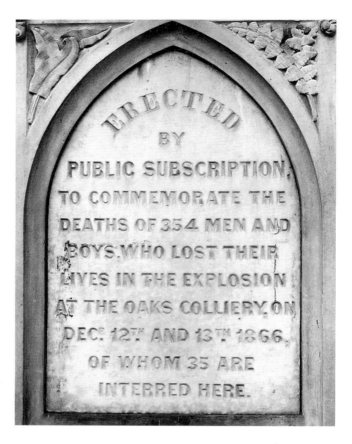

103 *Detail of Barnsley Oaks Memorial. (AA93/275)*

effects of the blast killed 320 of the 340 men and boys who were working underground at the time and of the 20 rescued 14 later died of their injuries. The explosion was heard three miles away and was described by the *Barnsley Chronicle* as shaking 'the whole neighbourhood as if the earth had been rent by an earthquake' (Duckham 1976, 69).

Following the first explosion 198 rescuers descended into the pit, twenty-seven of whom were killed when the pit exploded again at 9.00 am the following morning. The *Yorkshire Post*, reporting the second blast, declared that 'The heart sickens in contemplating such a frightful addition to the deplorable record of this unparalleled catastrophe' (Duckham 1976, 77). By the time rescue operations ceased the death toll was 361 men and boys. It was the worst disaster in the annals of coalmining in England. (The worst disaster in Great Britain was at Senghenydd in Wales in 1913, which claimed 439 lives.) At the time of the disaster the Oaks Colliery was one of the largest pits in Yorkshire, having three shafts and some fifty miles of underground galleries, these being ventilated by two furnaces. Unlike the Hartley disaster of four years before, no legislation was rushed through Parliament to improve safety in the industry. It did, however, provide further stimulus for the development of improved methods of ventilating underground workings.

Mines Rescue Station, Station Road, Ashington, Northumberland

NZ 274 877

105 *Former Mines Rescue Station, Ashington, Northumberland. (AA93/1092)*

The provision of stations where mines rescue equipment could be stored was recommended by the Royal Commission on Accidents in Mines (1880–6). The Commission also reported in favour of the provision of facilities to enable men to be trained in the use of such equipment. A decade later W E Garforth, a pioneer in the field of mines safety, hoped that 'in the near future, groups of collieries will establish and support central-district depots for the storing of "colliery accident apparatus" and the training of men to explain and use them' (*Transactions of the Institution of Mining Engineers* **XIV**, 495). The first mines rescue station in England was built at Tankersley, South Yorkshire in 1902 [**104**]. At that time the provision of rescue stations

104 *Tankersley Mines Rescue Station, Tankersley, South Yorkshire. (AA93/1458)*

was still voluntary, a state of affairs that continued until the Coal Mines Act of 1911. This Act repealed all earlier safety legislation and replaced it with a more coherent system of regulation, one of the requirements of which was that rescue brigades be set up in every coalfield. By 1914 it was compulsory to have a rescue station within 10 miles of each coalmine (later increased to 25 miles).

The Ashington mines rescue station [**105**], opened in November 1913, was one of a number established by a consortium of Durham and Northumberland coal companies (others were at Crook, Elswick and Houghton le Spring). The station, now a shop, was built next to the town hall and doubled as the community's fire station. It has a plain neo-classical façade of three storeys and six bays, the end bays projecting slightly and having quoined returns. When first built it had a staff of six (housed on the upper floors) and garaging for an Armstrong Whitworth Rescue Car and a Merryweather fire engine (Harrison 1990, 62).

Training gallery, Mines Rescue Station, Dorman Avenue South, Aylesham, Kent

TR 2377 5199

106 Exterior of training gallery, former Mines Rescue Station, Aylesham, Kent. (BB93/8669)

Breathing apparatus for rescue work was slow to be adopted by English colliery companies. An early use was at Seaham in 1881, when rescuers wore Fleuss equipment to search the pit following an explosion. By the early 20th century a number of types of equipment were in use, including smoke hoods supplied with air through pipes and self-contained apparatus such as the Draegar, Aerophor and Proto equipment. Rescue and Aid Order (Statutory Rules and Orders, 1912, No. 347) laid down that members of rescue teams had to be given a course of training which included 'practices with breathing apparatus in a gallery so constructed as to represent the conditions existing in a roadway of a coalmine'. During practices men were to carry out a number of tasks in the gallery while wearing the equipment, these to include 'building and removing temporary stoppings. Removing debris in confined spaces ...' and '... carrying a dummy weighing 150 lbs on a stretcher along the length of the gallery' (*The*

108 *Stained glass medallion depicting a member of a rescue team wearing breathing apparatus, NUM offices, Barnsley, South Yorkshire. (AA93/2663)*

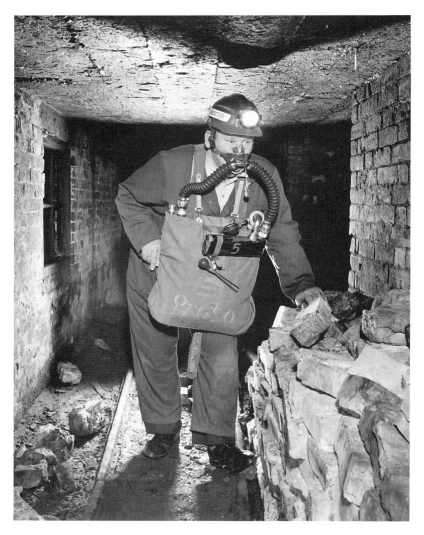

107 *Interior of training gallery, former Mines Rescue Station, Aylesham, Kent. (BB93/8671)*

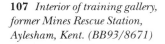

Colliery Guardian 10 May 1912). Mines rescue stations were provided with purpose-built training galleries, either under the buildings themselves or on land at the rear.

The mines rescue station at Aylesham, Kent [**106** and **107**], was shut down following the closure of Betteshanger Colliery in 1989. Since 1991 it has been used as a training establishment and is run by two of its former staff, Terrance Osborne and Graham Wilkinson, who provide training in rescue techniques and in the use of breathing apparatus.

Mines Research Station, Harpur Hill, Buxton, Derbyshire

SK 055 705

109 *Administration block, Mines Research Station, Harpur Hill, Buxton, Derbyshire. (AA93/1431)*

The dangers posed by the presence of firedamp were well understood by the early 19th century, but what had not been recognised was the potential of coal dust to combust under certain circumstances. In the 1840s Michael Faraday argued that coal dust had the effect of extending the flame of firedamp explosions. Almost half a century later, in 1891, a Royal Commission was appointed 'to enquire into the effect of coal dust in originating or extending explosions in mines, whether by itself or in conjunction with firedamp; and also to ensure whether there are any practical means of preventing or mitigating any dangers that may arise from the presence of coal dust in mines' (Roberts and King 1986, 6). The Commission reported that in its opinion coal could both cause and extend explosions, but there were those who questioned its findings. In 1906 another Royal Commission was appointed to look into a number of questions relating to the health and safety of miners, one of which was the still contentious issue of the dangers of coal dust. Between 1906 and 1908 a series of experiments was carried out at Altofts, West Yorkshire, these proving to all who witnessed them that coal dust could cause explosions, even in the absence of firedamp. In 1910 the government agreed to provide funding for continued research, and in May 1911 the Explosion in Mines Committee held its first meeting. A research station was established at Eskmeals in Cumbria, the buildings of which included a chemical laboratory and a Safety Lamp Testing Station.

The next major step forward was the creation of the Mining Dangers Research Board, created as a result of the Mining Industry Act of 1920. The purpose of the Board was 'to direct generally the work of research of the Mines Department into the causes of mining dangers and the means for preventing such dangers and to undertake the re-organisation of existing arrangements for carrying out such works at the Mines Department Experimental Station'. A year later the name of the body was changed to the Safety in Mines Research Board, under which name it continued to operate until Nationalisation. The Miners' Welfare Committee, another creation of the Act of 1920, agreed to provide funds for the Board's research work and, in particular, to finance a new experimental station.

The site chosen for the station was at Harpur Hill, near Buxton, Derbyshire. Experimental work was underway by the summer of 1926, a year before the station's official opening in June 1927. By 1935 the station had a number of explosion galleries, laboratories, a dining hall, accommodation for visiting staff, an exhibition hall and administration block [109]. Safety training was an important aspect of the station's work. In 1937, 6,153 people visited the station, and 110 lectures were given at other venues. When the coal industry was nationalised the research station at Harpur Hill became part of the Safety in Mines Research and Testing Branch, in the Health and Safety Branch of the Ministry of Fuel and Power. In 1975 the station became the largest single part of the newly created Health and Safety Executive's Research and Laboratory Services Division.

110 (opposite page) West Staithe, North Blyth, Northumberland. Work began on building the staithe in 1911 but it was not opened until 1928. It was built to ship coal produced by the Ashington Colliery Company's pits at Ashington, Bedlington, Ellington, Lynemouth and Woodhorn. The staithe ceased operation in December 1989. (BB93/28341)

6

Transport

111 Dunston Staithes, Gateshead, Tyne and Wear, photographed prior to restoration. (BB93/545)

Dunston Staithes, Gateshead, Tyne and Wear

NZ 189 525

By the early 17th century more than 200,000 tons of coal were being shipped from the River Tyne every year. It is estimated that by 1700 around two-thirds of the total output of British pits was produced at collieries close to the banks of the Tyne (Hill 1991). At that time it was usual to transport the coal downriver in open boats called keels, 'a broad flat-bottomed boat of oval form, designed to carry the heaviest load with the least draught of water' (Galloway 1969, 15). The coal was then transferred directly from the keels into the holds of seagoing vessels. However, towards the end of the 18th century some mine owners began to construct timber staithes which were linked to their pits at first by horse-drawn wagonways and later by steam railways. These staithes allowed coal to be carried from pit to ship without any need to use keels. The keelmen, recognising the threat this innovation posed to their livelihoods, took industrial action. A riot by Sunderland keelmen in 1815 resulted in the attempt to destroy one of the offending staithes, while a strike of Tyneside keelmen in 1822 led to pitched battles at Newcastle and North

Shields. The massive increase in coal production in the 19th century led to the building of ever larger staithes. Writing in the mid 19th century, J R Leifchild, noted that:

> It is a remarkable natural advantage, that the great northern coal-field is intersected by three considerable rivers, in consequence of which fact, the whole district possesses an easy, cheap, and expeditious mode by which its produce of coal may find its way into the general market. These three rivers are the Tyne, the Wear, and the Tees, each of which is admirably adapted by its volume of water, its tides, and harbour-room, for these purposes. The large collieries in the vicinity of the river, have each a running railway in the most direct line to the river banks. (Leifchild 1856, 75–6)

The Dunston Coal Staithes [111], built by the North Eastern Railway and completed in 1893, are one of the largest timber structures in the world. They were built in a period when coal production and coal exports were approaching their peak. British Rail ceased to use the staithes in 1983, but in 1990 they were restored for the Gateshead Garden Festival.

112 *New Holland Bulk Terminal, Immingham, Humberside. (AA93/1364)*

New Holland Bulk Terminal, Immingham, Humberside

TF 185 175

Since the mid 1970s Great Britain has imported more coal than it has exported. In 1991–2 the country imported 20,000,000 tons of coal and exported only 1,400,000 tons. The New Holland Bulk Terminal was built in 1983 to handle both imports and exports of coal and coke [112]. It also handles other cargoes, including feedstuffs, grain and fertilisers. The facility can take trains of up to 1,000 tons and has a capacity of 400 tons per hour.

Seaham Harbour, Seaham, County Durham

NZ 43 49

113 *North Dock, Seaham Harbour, Seaham, County Durham. (BB93/466)*

Of all the coal-shipping facilities built in the north east of England in the first half of the 19th century, the most ambitious was the Marquis of Londonderry's harbour at Seaham [**113**]. In the early 1820s Londonderry was making plans to build a railway and harbour to ship out coal from his wife's pits at Rainton, thereby saving himself the £10,000 a year that he paid keelmen to transport his coal down the River Wear from Penshaw staithes to the collier brigs moored down-stream. The foundation stone of Londonderry's new harbour at Seaham was laid on 28 November 1828 and on 31 July 1831 the first coal was transported down to the harbour on the newly completed Rainton and Seaham Railway and loaded on to the brig *Lord Seaham*. This event was the occasion for an inauguration ceremony, accompanied by bands, banners and speeches. Soon after, a wagonway was built from Seaham to a new pit at South Hetton, the first coal being shipped from North Dock in 1833 (McNee and Angus 1985, 5).

Seaham Harbour was extended a number of times in the course of the 19th century, the first addition being the South Dock which was opened in July 1835. In 1837, 1,782 vessels used the harbour and 370,000 tons of coal were carried down to it along the railways from Rainton and South Hetton. The new harbour, like its counterparts on the Tees and Wear, provided a considerable boost to the coal industry of County Durham. Leifchild wrote of Londonderry's port that 'Seaham harbour itself has in reality arisen out of coal, though in appearance it rises out of the sea' (Leifchild 1856, 79). However, by the middle of the century the coal trade and the ships which served it were both beginning to outgrow small harbours such as Seaham. In the 1850s Londonderry built a railway from Seaham to Sunderland in order to be able to ship coal out through George Hudson's newly built South Dock at the mouth of the River Wear.

A feature of Seaham Harbour was the unusual type of coal drop used to transfer coal from the wagons to the waiting colliers. The wagons were run on to platforms, which were then swung outwards and downwards to a position immediately above the ships' holds by means of a pivoting arm with a counterbalance. In 1966 the last drops at the South Dock were dismantled and in the 1970s were moved to the North of England Open Air Museum at Beamish and put into store. Consideration is now being given to returning the drops to Seaham and re-erecting them in the South Dock.

Bridgewater Canal Basin, Worsley, Greater Manchester

SD 005 747

In March 1759 Francis Egerton, third Duke of Bridgewater, obtained an Act of Parliament to build a canal from his collieries at Worsley to the growing town of Manchester. Work began in the same year and was completed in 1764 with the opening of the Castlefields basin at the Manchester end of the canal. The Bridgewater Canal was one of the first dead-water canals in the country [**114**]. The engineer appointed to oversee the work was James Brindley, a millwright experienced in the surveying of water courses. A feature of the Worsley mines was the fact that boats entered the mine workings by a network of tunnels and were loaded with coal underground. Work on the first of these tunnels, 'the Navigable Level', began in 1759 and the first workable coal, the 'Four Feet Seam', was reached in 1761. The canal proved a commercial success, leading the Duke to declare that 'a navigation should always have coal at the heels of it' (Lewis 1971, 20).

114 Bridgewater Canal Basin, Worsley, Greater Manchester (AA93/1385). The brick and timber-framed building was built for the first Earl of Ellesmere in the mid 19th century, replacing the original packet house of 1760.

It is significant that ninety of the 165 Acts for canal construction passed between 1758 and 1801 had the transportation of coal as their principal purpose (Lewis 1971, 20).

An immediate effect of the building of the Bridgewater Canal was that the coal brought to Manchester along it was sold at 6s per ton, whereas that brought in by river cost 12s. By 1800 more than 140,000 tons of coal a year were being transported on the canal. However, by 1846 the canal was proving inadequate to meet the needs of the mines it served, the general manager observing that 'our levels are nearly choked up with boats, and indeed, we are short of boats' (Church 1986, 43). The canal was also beginning to face serious competition from the railways. In 1864 the London and North Western Railway opened a line which ran through Worsley, after which an increasing amount of coal from the Bridgewater pits was carried by rail. On Nationalisation only three of the Bridgewater pits were still producing coal, while four others remained in use as pumping stations. The main Navigable Level and several branches were maintained for drainage. The last pit, Mosley Common, closed in 1968 and the remaining pumping stations on the old Level were abandoned.

Port of Goole, Goole, Humberside

SE 746 232

In 1820 the Aire and Calder Navigation secured an Act of Parliament for the building of a branch canal to the River Ouse and for the construction of two docks at Goole. Canal and port opened in 1826. Goole was well placed for transhipping coal from canal boats to seagoing colliers. However, by the 1850s the advantages of this shipping route had been undermined by competition from the railways and it became evident that speed of shipment would have to be increased and costs reduced if the canal were to continue carrying coal. The solution was devised by the Aire and Calder's engineer, William H Bartholemew, who came up with the idea of transporting coal in trains of compartment boats ('Tom Puddings') pulled by steam tugs [115]. At Goole several hydraulic hoists were erected [116], the purpose of which was to lift the compartment boats – each of which carried around 25 tons of coal – out of the water and tip their contents into seagoing colliers. The hoists, the first of which were erected in 1863, were designed by Bartholemew and used

115 *Train of compartment boats on the Aire and Calder Navigation at Castleford, West Yorkshire. (AA93/1050)*

116 *'Tom Pudding' hoist, Aldam Dock, Goole, Humberside. (BB93/7880)*

hydraulic equipment supplied by Sir William Armstrong's Elswick engineering works at Newcastle-upon-Tyne. Use of the compartment boat trains began in 1865, in which year they carried 9,145 tons of coal. Bartholemew's system proved a success, with traffic increasing steadily for the next fifty years, reaching 1,297,226 tons in 1910. Five hoists were built in all, only two of which now survive: the one in Aldam Dock (built 1889) and one in South Dock (1910). These were operating until the 1980s, but are now disused.

Coal is still transported on the Aire and Calder by the firm of Cawood Hargreaves, who use compartment boats to convey coal from Kellingley Colliery, North Yorkshire, to Ferrybridge Power Station, West Yorkshire.

Causey Arch, Stanley, County Durham

NZ 201 559

Wagonways, or Newcastle wagon roads as they were often called, were becoming common in the north east of England by the late 17th century. Roger North, writing in 1676, described the advantages of the wagonway:

> The manner of the carriage is by laying rails of timber from the colliery down to the river, exactly straight and parallel, and bulky carts are made with four rowlets fitting these rails, whereby the carriage is so easy that one horse will draw down four or five chaldrons of coals, and is an immense benefit to coal merchants. (Burton 1975, 45)

The Causey Arch was built in 1727 to carry a branch of the Tanfield Wagonway over the valley of the Bobgins Burn [**117**]. It is a stone bridge with a single masonry arch of 105 ft in length, 80 ft in height, and is claimed to be the oldest surviving railway bridge in the world. It has been estimated that it cost £2,250 to construct (Mann 1984, 224). The bridge was built for Captain Liddell and Mr Wortley as part of a wagonway system that ran south from the River Tyne, the purpose of which was to allow the working of coal in the Pontop and Tanfield areas.

In 1726 the Liddell and Wortley families joined with George Bowes, William Cotesworth and William Ord to establish a partnership of colliery owners known as the 'Grand Alliance'. By the middle of the century the Alliance owned 60 per cent of the sea-sale collieries south of the Tyne. The Liddells later became Lords Ravensworth and the Bowes, Lords Strathmore. In the late 18th century the introduction of steam pumps to make workable the coal on the banks of the Tyne east of Newcastle led to a depression in the coalfield west and south of the town. As a result the wagonway went into decline, and the Causey Arch was disused by the early 1800s. In 1812 the arch was described as 'at present neglected and falling into ruins' (Mann 1984, 224). Parts of the former Tanfield Wagonway are now a preserved railway.

Bowes Railway, Washington, Tyne and Wear

NZ 282 580

By the end of the 18th century the traditional wooden tracked wagonway was being superseded by the cast-iron plateway and, increasingly, by the railway. The next major step forward was the building of steam locomotives to replace horses as the principal means of motive power. It was in this important area that a number of colliery companies in the north of England made an important contribution to the development of the railway. The first successful use of steam locomotives was on the Middleton colliery's railway, near Leeds, West Yorkshire, in 1812. The engines, which worked on the rack and pinion system, were designed by the colliery's viewer, John Blenkinsop, and built by the Leeds firm of Fenton, Murray and Wood. By the end of 1813 steam locomotives of the type devised by Blenkinsop were working on the Orrel Colliery Railway, near Wigan, and on the Kenton and Coxlodge Collieries Railway, on Tyneside. In that year William Hedley experimented with steam locomotion on the Wylam Colliery Railway in County Durham. A year later George Stephenson took his first step towards becoming a railway engineer by building a locomotive for the Killingworth Colliery, also in County Durham.

The Bowes Railway was one of a number of colliery lines to be constructed in County Durham in the first half of the 19th century [118]. It was built for the firm of Lord Ravensworth and Partners, George Stephenson acting as engineer. In its original form the railway ran from staithes on the Tyne at Jarrow to Mount Moor Colliery, by way of Springwell Colliery. The first four miles from Jarrow were locomotive hauled, while the last two and a quarter were rope-hauled inclines. The locomotives were ordered from the recently established engineering firm of Robert Stephenson and Company and were the first 'travelling engines' to be built by it (the third was for the Stockton and Darlington Railway). The railway opened as far as Springwell in January 1826, horses being used on the level section because of delays in delivering the two steam locomotives ordered. In May 1842 an extension of the line to Kibblesworth Colliery was opened, the construction work involving two more inclines. By 1851 the railway had been taken over by John Bowes and Partners, a firm which owed much to the money and energy of the Tyneside shipbuilder, Charles Mark Palmer. In 1850 Palmer opened a shipyard at Jarrow and in the following year launched the first commercially successful screw-driven steam collier, the *John Bowes*. It was Palmer who, in 1853, gave the partnership's railway system the title of Pontop and Jarrow Railway, which it kept until renamed the Bowes Railway in 1932. At its peak the railway carried over 1,000,000 tons of coal a year.

In 1968 the system began to contract and by the end of 1974 only the short section from Monkton Coking Plant to Jarrow Staithe remained in use. The coke plant and staithe have since closed, the former being demolished in 1992 and the latter in 1993. In 1976 the importance of the railway – one of the oldest in the world and a rare example of the continued use of rope-hauled inclines – was recognised and steps taken to preserve at least part of it. In that year the former Tyne and Wear County Council purchased the one and a quarter mile stretch of the railway from Black Fell Bank Head to Springwell Head. Restoration and maintenance work has been carried out by volunteers belonging to the Tyne and Wear Industrial Monuments Trust and by staff employed under the Manpower Services Commission job creation schemes.

119 *Eastern Coal Drops, King's Cross, Camden, Greater London. (BB93/5514)*

Eastern Coal Drops, King's Cross, Camden, Greater London

TQ 301 837

The first coal to arrive in London by rail was transported to the capital by the London and Birmingham Railway in 1845 (Hunter and Thorne 1990, 112). In the thirty years that followed the amount of coal transported by rail increased and by 1870 more was arriving in London by train than in ships. The relative cheapness of coal carried by rail in this period was partly the result of a price-cutting war between the Great Northern and Midland Railway companies, both of whom established large coalyards at King's Cross and St Pancras. This opening up of the London market at competitive freight rates encouraged the development of the East Midlands and Yorkshire coalfield in particular. (In 1870 some 50 per cent of railborne coal arriving in London was from the former coalfield and 20 per cent from the latter (Church 1986, 45).) In 1880, 6,196,000 tons of coal came to London by rail, compared with only 4,000,000 tons by sea.

The eastern coal drops at King's Cross were built by the Great Northern Railway and brought into use in 1851 [**119**]. At the time of their opening it was claimed that they would be capable of handling 1,000 tons a day. The drops are brick built with arched openings in the side walls and a gabled roof, much of which was burnt off by a fire in 1985. The full coal trucks entered the structure at the upper level and discharged their loads into hoppers, which in turn discharged into a bagging-up area below.

Eggborough Power Station, Eggborough, North Yorkshire

SE 665 270

120 *Coal train approaching Eggborough Power Station, North Yorkshire. (AA93/1450)*

Today, rail is still the principal means of moving coal from pits and import terminals to the end users, most of which are power stations [**120**]. In 1913 around 3 per cent of coal produced in Great Britain was used to generate electric power. At Nationalisation in 1947, power stations still accounted for only 14 per cent of coal used in Great Britain, but in 1991–2 they were consuming around 65 per cent of home-produced and imported coal. In 1991–2 power stations consumed 80,500,000 tons of coal, 73,000,000 tons of which were produced by British Coal. Coal trains deliver approximately 1,100,000 tons of coal to power stations every week, this being carried by 1,200 trains operating on a twenty-four hour, six-days-a-week schedule (Anon 1992, 10).

Bibliography

Anon nd. *Three Men*. International Miners Mission

    1910. *Architectural Review* (Town Planning and Housing Supplement No. 3), March 1910

    1955. *Brodsworth Main Colliery, 1905–1955: Jubilee Souvenir*

    1990. *The Compleat Collier*. (First published 1708. References are to 1990 reprint, Dicks Publishing, Wigan)

    1992. *British Coal Corporation: Report and Accounts 1991/92*. British Coal

Ashworth, W 1986. *The History of the British Coal Industry* **5** *1946–1982: The Nationalised Industry*. Clarendon Press

Atkinson, F 1968. *The Great Northern Coalfield*. University Tutorial Press, London

Boulton, W S 1908. *Practical Coal Mining*. Gresham, London

Bucknell, L H 1935. *Industrial Architecture*. The Studio, London, 193–201

Burton, A 1975. *Remains of a Revolution*. Sphere Books, London

Caffyn, L 1986. *Workers' Housing in West Yorkshire, 1750–1920*. HMSO

Challinor, R and Ripley, B 1968. *The Miners' Association: A Trade Union in the Age of the Chartists*. Lawrence and Wishart, London

Church, R 1986. *The History of the British Coal Industry* **3** *1830–1913: Victorian Pre-eminence*. Clarendon Press

Clayton, A K 1962–3. 'The Newcomen Engine at Elsecar, West Riding'. *Transactions of the Newcomen Society* **35**, 97–108

Coal and the Environment

Coal: British Mining in Art 1680–1980. (Exhibition catalogue, Arts Council of Great Britain)

Coalite, 75 Anniversary, 1917–1992

Colls, R 1987. *The Pitmen of the Northern Coalfield*. Manchester University Press

Cornwell, J 1991. 'Excavation and conservation works at the Golden Valley Colliery, Bitton'. *Bristol Industrial Archaeological Society Journal* **23**

Davies, J and Powell, R 1987. *A Green and Pleasant Land*. Cornerstone, Manchester

Davison, J 1973. *Northumberland Miners*. NUM (Northumberland Area), Newcastle upon Tyne

Defoe, D 1983. *A Tour Through the Whole Island of Great Britain* **3** (edition first published 1928. 1983 reprint, The Folio Society, London. First published 1721)

Dickens, C 1969. *Hard Times* (edition first published 1854. 1969 reprint, Penguin)

Duckham, Baron F 1976. 'The Oaks Disaster, 1866'. *Studies in the Yorkshire Coal Industry*. Manchester Univ Press

Emery N 1990. 'Corrugated iron public buildings in County Durham'. *Durham Archaeological Journal* **6**, 59–73

    1992*a*. *The Coalminers of Durham*. Alan Sutton, Stroud

    1992*b*. 'The Esh Winning Miners Memorial Hall'. *Durham Archaeological Journal* **8**, 87–94

Falconer, K 1980. *Guide to England's Industrial Heritage*. Holmes and Meier Publishers, Inc, New York

Flinn, M W 1984. *The History of the British Coal Industry* **2** *1700–1830: The Industrial Revolution*. Clarendon Press

Foster, C (ed) 1988. *North Eastern Record* **1**. Historical Model Railway Society

Frank, R 1991. *Photofile*. (First published 1983 France. 1991 edition, Thames and Hudson)

Fynes, R 1873. *The Miners of Northumberland and Durham* (reprinted 1985 by A V Petrie, Whitley Bay)

Galloway, E 1830. *History and Progress of the Steam Engine*

Galloway, R 1969. *A History of Coal Mining in Great Britain*. (First published 1882. 1969 edition, Macmillan and Co)

    1971. *Annals of Coal Mining and the Coal Trade*. (First published 1898 by *The Colliery Guardian*. 1971 edition, David and Charles, Newton Abbot)

Granville, A B 1971. *Spas of England* **1** *The North*. (First published 1841. 1971 edition, Adams and Dart, Bath)

Griffin, A R 1977. *The British Coalmining Industry*. Moorland, Buxton

Haigh, B 1989. *Bolsover Colliery: A Centenary History*. British Coal

Harrison, B 1990. *Ashington: A History in Photographs*. Northumberland County Library

Hart, C E 1953. *The Free Miners.* British Publishing Company, Gloucester

Hill, A 1991. *Coal Mining: a technological chronology, 1700–1950.* British Mining Supplement

Hinsley, F B 1972. 'The development of coal mine ventilation in Great Britain up to the end of the nineteenth century'. *Transactions of the Newcomen Society* **XLII** 1969–70, 25–39

Hobsbawm, E 1969. *Industry and Empire.* Pelican Books

Hobsbawm, E (ed) 1971. *Primitive Rebels.* (First published 1959. 1971 edition, Pitman Press, Bath)

Holland, J 1841. *The History and Description of Fossil Fuel, The Collieries and Coal Trade of Great Britain.* London

Hunter, M and Thorne, R 1990. *Change at King's Cross: From 1800 to the present.* Historical Publications, London

John, A V 1984. *Coalmining Women: Victorian lives and campaigns.* Cambridge University Press, Cambridge

Kismaric, S 1990. *British Photography from the Thatcher Years.* The Museum of Modern Art, New York

Leifchild, J R 1856. *Our Coal and Our Coal Pits.* Longman, Brown, Green and Longmans, London

Leland, J 1964. *The Itinerary* **4**. Southern Illinois University Press, Carbondale

Lewis, B 1971. *Coal Mining in the Eighteenth and Nineteenth Centuries.* Longman, London

Linsley, S M 1976. Industrial Archaeology of Electricity around Tyne and Wear. *Proceedings of the Fourth Institution of Electrical Engineers' Meeting on the History of Electrical Engineering.* Durham

McCall, B 1971. 'Beehive coke ovens at Whinfield, County Durham'. *Industrial Archaeology* **8**, 52–62

McCutcheon, J E 1963. *The Hartley Colliery Disaster.* McCutcheon

McNee, T and Angus, D 1985. *Seaham Harbour: The first 100 years 1828–1928*

Mann, J 1984. 'Causey Arch – A Note'. *Archaeologia Aeliana*, 5th ser, **XII**

Mee, G 1976. 'Employer: employee relationships in the Industrial Revolution: The Fitzwilliam Collieries'. In Pollard and Holmes (eds) 1976

Miners Welfare Fund Annual Report 1936, 1937

Mining People. HMSO 1945

NCB 1950. *Plan for Coal*

NUM Souvenir Brochure 6

Orwell, G 1937. *The Road to Wigan Pier.* Penguin

Parliamentary Papers 1842, 217–19

Percy, C M 1886. *The Mechanical Engineering of Collieries* **II**, 2nd edition. *The Colliery Guardian* office, London
   1888. *The Mechanical Engineering of Collieries* **I**, 2nd edition, *The Colliery Guardian* office, London

Pollard, S and Holmes, C (eds) 1976. *Essays in the Economic and Social History of South Yorkshire.* South Yorkshire County Council, Barnsley

Preece, G 1988. Pithead Baths and the Miners Welfare Fund. Unpublished MA thesis, Manchester Polytechnic

Preece, G and Ellis, P 1981. *Coalmining: a Handbook to the History of Coalmining Gallery, Salford Museum of Mining.* City of Salford Cultural Services

Report of the Royal Commission on the Coal Industry **I**. HMSO 1926

Roberts, A F and King, J E 1986. *A History of Mine Safety Research.* HMSO

Supple, B 1987. *The History of the British Coal Industry* **4** *1913–1946: The Political Economy of Decline.* Clarendon Press

The Colliery Guardian. London

The Iron and Coal Trades Review. 1927

The Law Relating to Mines under the Coal Mines Act, 1911. HMSO London 1914

Thompson, E P 1968. *The Making of the English Working Class.* (First published 1963. 1968 edition, Penguin Books)

Threlkeld, J 1989. *PITS 2.* Wharncliffe, Barnsley

Transactions of the Institution of Mining Engineers **XIV**, **XLVIII**

Tremenheere, H S 1845. *Report on the Mining Population in Parts of Scotland and Yorkshire*, Parliamentary papers **XXVII**, 25

VCH (Victoria County History) 1912. Yorkshire

Waller, R J 1979. 'A Company Village in the New Dukeries Coalfield: New Ollerton, 1918–39'. *Transactions of the Thoroton Society* **LXXXIII**, 70–9

Watkins, G M 1955. 'The Vertical Winding Engines of Durham'. *Transactions of the Newcomen Society* **XXIX**, 205–19
   1979. *The Steam Engine in Industry* **2**. Moorland, Ashbourne

Whitelock, G C H nd. *250 Years in Coal: The History of Barber Walker and Company, 1680–1946*

Yorkshire Mining Museum 1992. *Pit Ponies*

Index

(Page numbers in **bold** refer to illustrations)

Textile Mills

The three volumes on textile mills published by the Royal Commission
form a landmark in British industrial archaeology publishing. Together
they provide the most complete account so far of these buildings, which
play such a crucial role in our national heritage. Each book includes a
gazetteer of mills as well as chapters tracing the histories both of the mills
themselves and of the technologies that developed with them. They will
be of particular interest to historians, industrial archaeologists and those
concerned with the planning issues raised by industrial buildings.

EAST CHESHIRE TEXTILE MILLS

Anthony Calladine and Jean Fricker

East Cheshire Textile Mills is the first book to survey all the mills of this area,
the centre of the silk industry in England. The authors have established
that the factory system, and the architecture associated with it, in the
towns of Macclesfield and Congleton pre-dated by a generation the more
widely known factories of Richard Arkwright and others. The book goes
beyond the factories themselves and looks at the houses, landscapes and
communities generated by the industry.
*At last mills are receiving the sort of attention that churches
and country houses have long enjoyed.* **Country Life**
1 873592 13 2   £14.95

COTTON MILLS IN GREATER MANCHESTER

Mike Williams with D A Farnie

For 150 years, cotton dominated the economy of Greater Manchester.
This book, resulting from a joint project by the RCHME and the Greater
Manchester Archaeological Unit, traces the development of the cotton
mills in the area, from the early days of mechanisation through to the
industry's demise in the early 20th century.
*At last we have a proper, memory jerking record ... superb ... lavishly illustrated ...
a comprehensive archive of the mills.* **Bolton Evening News**
0 948789 89 1   £14.95

YORKSHIRE TEXTILE MILLS 1770–1930

Colum Giles and Ian H Goodall

The first volume to be published in the series, *Yorkshire Textile Mills* tells
the story of one of England's greatest industries. The architecture of the
area ranges from small, water-powered mills in the Yorkshire valleys to the
cathedral-like structures dominating such towns as Bradford and Halifax.
The mills are also seen in the context of the landscape and the communities that grew up around them.
*An excellent book ... both for research and general reading, this is
a first-class work.* **Yorkshire on Sunday**
0 11 300038 3   £16.95

AN ARCHITECTURAL SURVEY OF URBAN DEVELOPMENT CORPORATION AREAS

This series of illustrated ring-bound reports records the industrial archaeology of Urban Development Corporations (UDCs), areas established by Government in the 1980s and earmarked for redevelopment. The survey is aimed at all interested in the issues of urban redevelopment.

Bristol	0 9514896 7 4	£5.50
The Black Country	0 9514896 8 2	£7.50
Leeds	0 9514896 3 1	£5
Sheffield	0 9514896 4 X	£3
Teesside	0 9514896 6 6	£6
Tyne and Wear I (Tyneside)	0 9514896 5 8	£5
Tyne and Wear II (Wearside)	1 873592 14 0	£5

POTWORKS

The Industrial Architecture of the Staffordshire Potteries

The six towns of the Staffordshire Potteries – Tunstall, Burslem, Hanley, Stoke, Fenton and Longton – owe their popular name and distinctive character to one product: pottery. Though the development of the Staffordshire ceramics industry has been well charted, this book presents the first comprehensive analysis of the potworks themselves. *Potworks* traces the evolution of ceramic production from the early cottage-industry kilns of the 16th and 17th centuries to the purpose-built but often architecturally idiosyncratic factories of the 19th century. The book also looks at the wide range of buildings associated with the industry – such as the housing of workers and managers and the civic and institutional buildings erected by philanthropic potwork owners – and in so doing draws a rounded picture of Potteries society.

1 873592 01 9   £9.95

**For an order form for these and other titles write to:
RCHME, Publications Section,
National Monuments Record Centre, Kemble Drive,
Swindon SN2 2GZ.
Also available from good bookshops**